FOCUS ON CIVILIZATIONS AND CULTURES

NATIVE AMERICANS

CULTURAL DIVERSITY, HEALTH ISSUES AND CHALLENGES

Focus on Civilizations and Cultures

Additional books in this series can be found on Nova's website under the Series tab.

Additional e-books in this series can be found on Nova's website under the e-book tab.

FOCUS ON CIVILIZATIONS AND CULTURES

NATIVE AMERICANS

CULTURAL DIVERSITY, HEALTH ISSUES AND CHALLENGES

JEROME MENDOZA
EDITOR

nova publishers
New York

Library of Congress Cataloging-in-Publication Data

Native Americans : cultural diversity, health issues and challenges / editor, Jerome Mendoza.
pages cm. -- (Focus on civilizations and cultures)
Summary: "The first chapter of this book traces the rise of the idea of Manifest Destiny, showing how it contributed to the historical traumatization experienced by Native American tribes during an initial period in U.S. history (1790-1890). Illustrations of renewal and revitalization among present-day Native Americans, and an interpretation of their wider significance for our contemporary world are examined. In the second chapter, the relatedness of First America Inhabitants is explored with Pacific Islanders by using autosomal genetic markers: the HLA alleles. HLA is the most polymorphic human genetic system and is most useful for comparing populations' relatedness. The ethnic groups of Pacific Islanders, First America inhabitants and other World Populations were used. The authors of the third chapter use longitudinal data to examine mediational mechanisms that account for the relationship between ethnic identity and self-esteem among American Indian youth. The authors of the last chapter of this book explore the effect of food assistance on American Indian women's food choices and the role of food assistance programs in improving food choices and diet quality in vulnerable populations"--From publisher's website.
Includes index.
ISBN: 978-1-63483-184-0 (hardcover)
1. Indians of North America--Social conditions. 2. Indians of North America--Social life and customs. 3. Cultural pluralism--North America. 4. Indians of North America--Health and hygiene. 5. Pacific Islanders--Health and hygiene. 6. Human genetics. 7. Indian youth--North America--Social conditions. 8. Indians of North America--Food. 9. Food relief--North America. I. Mendoza, Jerome.
E98.S67N385 2015
970.004'97--dc23
2015027294

Published by Nova Science Publishers, Inc. † New York

CONTENTS

PREFACE

Chapter 1 - This chapter traces the rise of the idea of Manifest Destiny, showing how it contributed to the historical traumatization experienced by Native American tribes during an initial period in U.S. history (1790-1890). Among the topics treated are the consequences of Jackson's Indian Removal Act (1830), the development of 19th century theories of race, and the systematic attack on Native American cultural autonomy, including the Dawes Act (1887), the government supported boarding schools, and the systematic destruction of the bison herds. The paper concludes by providing both illustrations of renewal and revitalization among present day Native Americans, and an interpretation of their wider significance for our contemporary world.

Chapter 2 - America peopling has recently been explained based only on genetic data. While different First America inhabitants' ethnic groups, Amerindians, Na-Dene speakers, Aleuts and Eskimo there exist, there is no genetic, cultural or anthropological homogeneity within these groups. The authors have particularly addressed the relatedness of First America Inhabitants with Pacific Islanders by using autosomal genetic markers: the HLA alleles. HLA is the most polymorphic human genetic system and this is most useful for comparing population relatedness. Ethnic groups of Pacific Islanders, First America Inhabitants and other World Populations have been used. A genealogic study and also a frequency comparison studies by using HLA alleles and haplotypes have been carried out. The author's conclusions are: 1- Aleuts seem to be a genetic and linguistic separate group which may be related to northern European Lapps, both of them originated in southern Siberia Baikal Lake area. 2- First America Inhabitants, including all analyzed Amerindians, Na-Dene speakers and Eskimo have had genetic flow with Pacific Islanders:

the latter share autosomal HLA alleles and haplotypes with First America Inhabitants. This could have been bidirectional. 3- Particularly, Easter Islanders show a probable cultural and genetic exchange with Titikaka Lake Aymaras and a contact possibility according to Kon-tiki Lima/Polynesia Pacific expedition. This civilization also shares significant traits with European Iberian megalithic builders. 4- Mesoamericans may be grouped together because they bear more ancient Olmec culture traits and the author's HLA results. 5- Genetics is not able by itself to uncover in space and time America peopling and First American Inhabitants relatedness with Pacific Islanders and ancient Solutrean Europeans. 6- Megalithic and genetic data found in Azores Islands shows that a pre-Celtic/Iberian culture may have also reached these Islands (and may be America) in Iron/Bronze Neolithic periods. Thus, there are both genetic and cultural solid evidence for Trans-Atlantic contact between Europe and America in ancient times.

Chapter 3 - Although a strong ethnic identity has been associated with self-esteem, few studies have used longitudinal data to examine mediational mechanisms that account for this relationship, especially among American Indian youth. This study investigated religious identity as a mediator of the relationship between ethnic identity and self-esteem. Structural Equation Modeling was used to investigate mediation pathways from ethnic identity to self-esteem in a sample (N = 1,571) of Lumbee middle school students. Results showed that the relationship between Year 1 ethnic identity and Year 3 self-esteem was partially mediated through Year 2 religious identity. Implications were discussed.

Chapter 4 - Limited access to nutrient-rich foods in low-income neighborhoods as well as limited income resources may collectively result in poor diet quality. Federal food and nutrition assistance programs support one in four Americans in achieving adequate nutrition. The Special Supplemental Nutrition Program for Women, Infants, and Children (WIC) is one of the food assistance programs that plays an important role in improving food choices and diet quality in vulnerable populations due to its targeted impact and broad reach.

The WIC program serves participants by providing supplemental foods, nutrition education, breastfeeding support and referrals to health and social services. The goals are to improve birth outcomes, to support the growth and the development of infants and children, and the promotion of long-term health of all WIC participants. In 2009, the U.S. Department of Agriculture (USDA) instituted change in the types and the amount of foods offered to WIC participants. One of the main changes in WIC food packages includes the provision of vouchers for purchases of any eligible fresh, frozen, or canned fruits and vegetables. The revised WIC food packages were intended to improve the diets and health of culturally diverse WIC participants with a wide range of traditional food preferences. American Indians are a subgroup of the population that is served by the WIC program. Compared with the general U.S. population, the WIC population is distinctively poor, but American Indian WIC participants face additional risks as nearly 81.2% are at or below 130% of the poverty. Although it is documented that obesity, diabetes, heart disease and cancer are major problems affecting American Indians relatively little is known about their eating behaviors. Critical factors present within Indian reservations may limit the ability to access and consume fruits and vegetable as well as other healthy foods. This study addresses the effect of WIC food-package revisions on frequency of juices and frequency and variety of fruits and vegetables consumed by American Indian WIC women. There were modest changes in food choice after the revisions in the food package. Specifically, the consumption of vegetables and some fruits increased, while the consumption of juice declined in frequency.

In: Native Americans
Editor: Jerome Mendoza
ISBN: 978-1-63483-184-0

Chapter 1

MANIFEST DESTINY AND TRAUMATIZATION OF INDIGENOUS NATIVE AMERICAN CULTURE: SOUL WOUND FOLLOWED BY RENEWAL AND REVITALIZATION

Steven Borish[1], Arthur Scott[2], Mietek Kolipinski[3], Marie Pace[4] and Sibdas Ghosh[4,*]
[1]Department of Human Development, California State University East Bay, Hayward, California, US
[2]Department of Humanities, Dominican University of California, San Rafael, California, US
[3]National Park Service Pacific West Regional Office, San Francisco, California, US
[4]School of Arts and Science, Iona College, New Rochelle, New York, US

ABSTRACT

This chapter traces the rise of the idea of Manifest Destiny, showing how it contributed to the historical traumatization experienced by Native American tribes during an initial period in U.S. history (1790-1890). Among the topics treated are the consequences of Jackson's Indian Removal Act (1830), the development of 19th century theories of race,

[*]Corresponding author: SGhosh@iona.edu.

and the systematic attack on Native American cultural autonomy, including the Dawes Act (1887), the government supported boarding schools, and the systematic destruction of the bison herds. The paper concludes by providing both illustrations of renewal and revitalization among present day Native Americans, and an interpretation of their wider significance for our contemporary world.

TRAUMA AND THE MIRROR OF INTERCONNECTEDNESS

> We've been taught this from the beginning of our lives: take care of this land and everything that's on it; take care of it well in order to bring good to all the plant life and all the things that are here. We have to take care of them all. It is very important for us today, as I see it, to come back to the Native way of life. The Native way is to pray for everything.
>
> – Corbin Harney, Western Shoshone Medicine Man [1, p. 8].

An important part of the literature of trauma focuses on individual development through different stages of the human life cycle [2, 3]. A second perspective is provided by the literature focusing on individual trauma in the context of specific historical episodes followed over decades, such as the literature dealing with the short and long term trauma of the wars in Iraq and Afghanistan for all participants, including Iraqis, Afghans and U.S. veterans [4, 5]. A third perspective integrates and extends the perspective of the first two, dealing with the effects of long term historical traumatization, as in studies of child survivors of the Holocaust [6] and of the *hibakusha* (lit: "the people affected by the explosion), Japanese survivors of the atomic bomb explosion in Hiroshima in 1945 [7].

Relevant here as well are "the children of the white mist," a term referring to the epidemic of birth defects in Vietnam related to the use of Agent Orange in the military campaign labeled "Operation Ranch Hand" initiated by the U.S. Air Force. A meta-analysis published in the *International Journal of Epidemiology* estimates that between 1961 and 1971 "49.3 million liters of Agent Orange containing more than 360 kg dioxin-contaminated defoliants was sprayed multiple times over 2.6 million acres" [8]. Dioxin (2,3,7,8-T4CDD) is one of a group of toxic chemicals that function as "synthetic endocrine disruptors" with frequent severe consequences for development of the embryo in both human and non-human species [9].

Two important recent works that exemplify this third perspective on historical traumatization trace the corrosive stigmatization suffered by

African-Americans through the social construction of "black criminality" [10] and the replacement of the older Jim Crow with a new systems level policy of mass incarceration [11]. It may be helpful to conceptualize the effects of long term historical traumatization as a form of "soul wound" [12, 13]. Historical traumatization includes individual psychological trauma as a subset of events: the trauma(s) experienced by members of one or more generation becomes intergenerational trauma, passed down over many generations from each generation to the next as part of their unfortunate and in some cases tragic cultural inheritance. In this paper, building on the work of Braveheart [14, 15, 16] and others [12], we focus on the third of these three perspectives, limiting our discussion to Native-Americans in North America. We start our historical review with events in the early 17th century.

In the case of the North American Native Americans, the unfolding of their historical trauma began as early as the 17th century with the arrival of the Pilgrims in Plymouth in 1620. As two historians have written of this period: "The terrible truth of Puritan-Indian relations in seventeenth century New England is that Puritan acquaintance with Indians in economic, political and military matters did not lead to greater understanding and appreciation of native culture. Rather, it confirmed the image of self-satisfied barbarians who denied God's commandment to live exactly as the Puritans lived" [17, p. 38]. After the devastation of numerous epidemics (1616-19, 1633-34 and others after 1650), and two disastrous wars (the Pequot War, 1636-37 and King Phillips's War, 1675-76), Native Americans had faced for the first time the ultimatum they would face over and over again in the westward movement of American society: "sacrifice your own culture and adopt the life-style of Anglo-Americans, or die in the act of resistance" (Ibid.)

As we shall see later, even attempting to become like white Americans did not necessarily save them from further forced migration, humiliation and cultural loss [18]. The issue was always land: the Native-Americans had it, the white Americans wanted it. The more devout among the Puritans saw it differently. The preacher Cotton Mather made the argument that "the Devil decoyed these miserable savages [to New England] in hopes that the Gospel of the Lord Jesus Christ would never come here to destroy or disturb his absolute empire over them." His father, Increase Mather, went even further: "We may truly say of Philip, and the Indians, who have fought to dispossess us, of the land, which the Lord our God hath given to us, as sometimes Jepthah, and the Children of Israel said to the King of Ammon, I have not sinned against thee, but thou dost me wrong to war against me…" [17, pp.182-183). By this reasoning, Native Americans could be seen as agents of Satan; moreover, their land was not being taken from them, because in the mind of these Puritans it never belonged to them in the first place.

In later sections of this paper, we will argue first, that systematized multilevel invalidation is in evidence here, as in Bateson's [19] theory of the double-bind, and second, that the forms of intergenerational trauma set in motion in this 17th century example have lasted until the present day. As in any social-historical situation where there is systematic and deeply induced cultural trauma that passes down through generations, observers without the blinders who are not vested in serving the existing power structure can clearly see the harm done. To take an example which may seem far afield but is not: In late 18th century Europe Oluf L. Bang (1731-1789), a powerful attorney general defending the proposed Danish Land Reforms (1784-1797) that would successfully – and without violence - end a centuries old feudal system, wrote in 1785 the following lines which point to historically induced trauma among the Danish peasants:

> Where this exists, the Farmer is utterly oppressed, [and] an oppressed man can neither desire nor act well; the concern that all Created Beings have for their deeds, which is their own welfare, is lacking in him. He may do as he will, he sees nothing other than Poverty and Oppression; he may finally stop wanting his own Best [interests]; because he sees that he can't attain it. This despair is the first thought that each upcoming generation drinks in. It sets roots, propagates itself, and becomes a source of corruption to entire generations…I find this mode of treatment both completely destructive and completely unjust (cited in Borish [20], p. 131, original in Hansgaard [21], p. 111).

With these long ago words from Denmark echoing in our ears, let us begin with a brief review of health consequences among present day Native Americans. Across the board, the rate of health problems related to tobacco, illicit drugs, alcohol, obesity, diabetes and suicide for Indians and Alaska Natives far exceeds similar rates for the rest of the population. Taking just the issue of alcohol, Fetal Alcohol Syndrome is a condition that can result from continued alcohol abuse during pregnancy. The symptoms found in a child with Fetal Alcohol Syndrome include facial anomalies, heart defects, low birth weight, behavioral problems, and mental retardation, as well as poor physical coordination and seizures. Fetal Alcohol Syndrome can also cause stillbirth and premature labor [22]; for a personal account of the adoption of an Indian boy suffering from Fetal Alcohol syndrome [23]. Other injuries related to alcohol and substance abuse can include chronic liver disease and cirrhosis, suicide and homicide.

Native Americans are twice as likely to commit suicide, with 75% of these suicides being alcohol-related, i.e., "depressive symptoms exacerbated by continued heavy drinking" [24, 25]. According to the Indian Health Service, American Indians in the 12 Indian Health Service areas "have higher rates of death from tuberculosis (500% higher) alcohol (514%), diabetes (171%), unintentional injuries (140%), homicide (92%), and suicide (82%) than all other US ethnic and racial groups" [26]. Native Americans in both urban and rural America are twice as likely as the general population in these areas to be poor, employed and without a college degree [27]. The psychological impact of the humiliating residential schools to which many children were subjected after being forcefully extracted from their families of origin is an additional contributing factor [13, 28].

A final point to consider pertains to being part of a people whose cultural memory has been, and is, subjected to continual, ongoing invalidation. As an example: on April 22, 1989 "Oklahomans on horseback and in wagons...reenacted the mad rush that occurred on this date a century ago when two million acres of Indian territory were opened to settlement by whites" (Land Rush Re-enacted With Yells: *New York Times,* April 23, 1989, cited in Borish[20], p. 61). A century earlier - on April 22 1889 - 40,000 settlers had raced to claim their 160 acre lots on land (*Indian land* in Oklahoma) that had just been declared open to settlement by President Benjamin Harrison. A member of the Tribal Voices coalition responded to the yells and good cheer of the celebration with this trenchant comment: asking Indians to take part in this centennial event was like "asking us to dance on the graves of our people."

In the following section, the events of this historical context are developed in a deeper perspective based on social and anthropological analysis. We conclude this section with two representative quotations from Native Americans, allowing them to illustrate their view of the historical record:

> We did not think of the great open plains, the beautiful rolling hills, and winding streams with tangled growth as wild. Only to the white man was nature a wilderness and only to him was the land infested with wild animals and savage people. To us it was tame. Earth was bountiful, and we were surrounded with the blessings of the Great Mystery. Not until the hairy man from the east came, and with brutal frenzy heaped injustices upon us and the families we loved was it wild for us. When the very animals of the forest began fleeing from his approach, then it was for

us that the Wild West began (Luther Standing Bear, Sioux Chief, cited in Curtis and Brown [29], p. 86)

What treaty that the whites have kept has the red man broken? Not one. What treaty that the white man ever made with us have they kept? Not one. When I was a boy the Sioux owned the world; the sun rose and set on their land; they sent ten thousand men to battle. Where are the warriors today? Who slew them? Where are our lands? Who owns them?.... Is it wrong for me to love my own? Is it wicked for me because my skin is red? Because I am Lakota, because I was born where my father died, because I would die for my people and my country? (Sitting Bull (Lakota), cited in Matthiessen [30], p. 33).

Manifest Destiny and the War on Native-American Cultural Autonomy

They are taking us beyond Miami
They are taking us beyond the Caloose River
They are taking us to the end of our tribe….
--from a Seminole song, collected by Frances Densmore (1867-1957), cited in Bruchac [18].

Manifest Destiny and Turner's Frontier Thesis

The term Manifest Destiny was coined by the influential columnist and the editor John L. O'Sullivan in an essay entitled "Annexation" (1845), published in the *United States Magazine and Democratic Review*. In his essay he urged the government to annex the Republic of Texas because it was "our manifest destiny to overspread the continent allotted by Providence for the free development of our yearly multiplying millions" (cited in Hietala [31], p. 255). The historian William Weeks [32] describes three central themes typically found in the claims for Manifest Destiny: virtue (the idea of the unique moral goodness of the American people and their institutions); mission (the responsibility to spread these institutions); and destiny under God (the divine support ensuring success in this work). Of course the expansionist idea behind the phrase "Manifest Destiny" long predated 1845; for example, in 1814 Matthew Lyon, a member of Congress from Vermont, wrote a letter to Andrew Jackson with these words: "This Nation are *[sic]* destined to civilize and Govern this Continent" [33].

The following discussion will focus on three critical periods of U.S. History strongly influenced by the belief in Manifest Destiny and will trace its devastating consequences for Native Americans. We will discuss first, *the period of Jeffersonian expansion* under the three "Jeffersonian Presidents" (Jefferson, Madison and Monroe, 1790-1825); second, *the Jacksonian period with a focus on the Indian Removal Act* (1830) and the late Jacksonian era (1825-1850); and third, *the period of the final westward expansion* (1850's to 1890), the chaotic period of the post-midcentury "Indian Wars," a time of major historical transformations. These include the Civil War, the coming of the railroads, the Gold Rush in California and the Black Hills, the destruction of the bison herds, and the massive overland migrations of white immigrants which accompanied the final movement of the American frontier to California and the Pacific Ocean. It is during the westward expansion in these three periods of nineteenth century history that the traumatic loss of Native American cultural autonomy escalated with tragic consequences: deculturation, loss of sacred lands, forced imprisonment on reservations, impoverishment, disease, and stigmatization based on "the scientific study of race" that was a dominant 19th century American paradigm. It is important to point out that the harm done to Native Americans was not an accident but a specific goal of U.S. policy, such as the mass slaughter and destruction of the bison herds that once roamed North America [34, 35].

Viewed as a whole, the impact of this period of American history is of such central importance that it requires special conceptual recognition. The historian Frederick Jackson Turner, writing in 1893, closed his insightful and classic essay "The Significance of the Frontier" with these words: "What the Mediterranean Sea was to the Greeks, breaking the bond of custom, offering new experiences, calling out new institutions and activities, that, and more, the ever retreating frontier has been to the United States directly, and to the nations of Europe more remotely. And now, four centuries from the discovery of America, at the end of a hundred years of life under the Constitution, the frontier has gone, and with its going has closed the first period of American history (cited in Turner [36], p. 59-60).

Turner [36] is eloquent, but even the wisest historian can be a prisoner of his own time and culture: America may have been "reached" by Europeans four centuries before Jackson's time, but it was clearly not "discovered" by Spanish and Italian sea captains. If we want to use the term "discover," America was most likely discovered by migrant hunters who crossed the Bering Straits land bridge some fifteen thousand years earlier. Turner was both correct and insightful in arguing that the first period of American history had just closed at the time he wrote his essay, but from the vantage point of 125 years later his assertion that "the frontier has gone" requires rethinking. It may be more correct to say that the frontier in a deeper sense has gone underground and-- spreading its threads of "manifest destiny" like some vast mycellial network -- has not only survived, but emerged as a key feature of the predominant cultural mythology that presently rules the North American political landscape. We suggest that a better term for this period of American history than the opening and closing of the frontier (1790 – 1890), one which more accurately reflects the experience of indigenous Native Americans, is "The Hundred Years Expansion." We will have cause to return to Turner's frontier thesis in this paper's conclusion.

The War Begins: Jeffersonian Expansion and the Legacy of Andrew Jackson

The first American figure who played a decisive role in the Hundred Years Expansion was Thomas Jefferson. When Thomas Jefferson assumed the office of the Presidency in 1801, his views on expansion differed considerably from his two predecessors George Washington and John Adams. As early as 1786 Thomas Jefferson had written: "Our continent must be viewed as the nest from which all America, North and South, is to be peopled." In 1801, the same year that he assumed the Presidency, he wrote to Governor James Monroe, "It is impossible not to look forward to distant times, when our rapid multiplication will expand itself...and cover the whole northern, if not the southern continent" (citations in Owsley and Smith [37], p. 16).

From the first European contact with Native Americans early in the 16th century, the North American continent had become a battleground for competing European colonial powers. England, France, Spain and smaller powers carved out territories and spheres of interest (usually at the expense of Native Americans). Yet by the end of the 18th century, the new republic of the United States had replaced France as a leader in the continental power struggle, and Jefferson was quick to take advantage of this new situation. In 1802 he authorized Robert R. Livingston and James Monroe to negotiate the purchase of territory east of the Mississippi River, including New Orleans. The settlement signed with France in 1803 allowed the United States to purchase the entire Louisiana region for $15 million dollars.

This action clearly shows Jefferson's strong support for westward expansion. And even while the negotiations were in progress, Jefferson secured funding for Meriwether Lewis and William Clark to explore the lands west of the Mississippi. Although in public he spoke of the mission's scientific purposes, Jefferson also wanted a military reconnaissance to evaluate the strength of Native Americans within the western territories, and an evaluation of the huge new region's suitability for American expansion. The Lewis and Clark expedition (1804-1806) "was but one of a series of incursions into Louisiana and Spanish territory under the guise of exploration" (citations in Owsley and Smith [37], p. 13).

While Jefferson's purchase of the Louisiana Territory can be found in every U.S. history textbook, the land expansion that took place along the Gulf Coast (1810-1820) culminating in the acquisition of Florida is much less well known. Under the three "Jeffersonian Presidents" (Jefferson, Madison and Monroe), expansionist policies were consistently followed, including waging Indian wars to strengthen U.S. claims to southern lands. In 1812 James Madison sent Andrew Jackson to prosecute Creeks hostile to the United States. Andrew Jackson's invasion of Florida during the spring of 1818, which took place under Monroe's Presidency, led directly to Spain's surrender of Florida to the United States. In the words of two historians of the period: "With the acquisition of Florida, southern Manifest Destiny was complete. All the lands east of the Mississippi River, in addition to the Louisiana Purchase, had been acquired without war. Afterward expansion focused westward, and within the next thirty years the nation would encompass the lands between the Atlantic and the Pacific..." (citations in Owsley and Smith [37], p. 15).

Although the westward expansion of the 13 states unquestionably began with Jefferson's Presidency, Jefferson himself seems to have maintained an appreciation and respect for Native American culture. In a letter written to John Adams in June 1812, he made these observations about the Cherokee, with whom he had visited as a young man: "That nation, consisting now of about 2,000 warriors, and the Creeks of about 3,000, are far advanced in civilization. They have good cabins, enclosed fields, large herds of cattle and hogs, spin and weave their own clothes of cotton, have smiths and other of the most necessary tradesman, write and read, are on the increase in numbers, and a branch of Cherokees is now instituting a regular representative government" (cited in Hartman [38], pp. 27-28). Even as late as the early 1820's, examples of mutual respect can be found in the historical record. Thus, the greatest of all Choctaw chiefs, Pushmataha (1760-1824) served in the War of 1812 for the U.S., negotiated several treaties with the U.S., and was buried in 1824 with full military honors in the Congressional Cemetery in Washington, D.C. (Anastasia [39], p. 29).

In the years after 1825, the screws began to tighten on Native Americans. What is called "the Doctrine of Discovery" allowed the United States (and all the major colonizing nations of the west) to simply disregard all claims made by indigenous people to the lands and territories that they had been occupying - in many cases for centuries. This opinion issued in 1823 by Chief Justice John Marshall of the U.S. Supreme Court in Johnson vs McIntosh "represents the most influential legal opinion of indigenous peoples' human rights ever issued by a court of law in the Western world" [Williams [40], p. 224). According to Marshall's frequently cited words in this case, European "discovery" of lands occupied by Native-Americans in the New World gave title "to the government by whose subjects, or by whose authority, it was made, against all other European governments, which title might be consummated by possession." A further finding of the court was that the discovering European nation had "an exclusive right to extinguish the Indian title of occupancy, either by purchase or conquest." As Williams [40, p. 224] points out, "this power of unilateral extinguishment of Indian rights to property and self-determination in tribal lands is the most important part of the discovery doctrine."

Marshall, who had served under George Washington at Valley Forge, and was responsible for securing the ratification of the Constitution from the key state of Virginia, was not a fringe figure; he was an insider's insider, a member of the Founding Generation. Under the Doctrine of Discovery as interpreted by Marshall, the sovereignty of the indigenous Americans was neatly and legally taken away from them; exclusive title now resided with their European "discoverers." The decades that followed would see other uses of law and the legal system to the same end: they would be used to deprive Native-Americans of their land, liberty and cultural identity.

In the same decade that Chief Justice Marshall laid down his Discovery decision, former general and Indian fighter Andrew Jackson began his two terms as President (1829-1837). Early in his first term of office, the desire of white settlers in the South to expand into lands belonging to five Indian tribes led Jackson to push through congress the Indian Removal Act (1830). After the passage of this Act, the U.S. government spent nearly 30 years forcing Native Americans to move westward, beyond the Mississippi River. The forced resettlement on "Indian land" in the west was not a true gift of new land to Native Americans, but only a temporary political solution. As noted by a historian of the period: "The Jacksonian Democrats knew that the western states and territories and the pioneers themselves would demand that Indian lands be turned over to the United States for use by white settlers – for when the Indians were promised land 'forever,' it meant not for all time, but only for as long as the territory remained unappealing to whites" (Hietala [31], pp. 136-137).

In one episode brought about by the Indian Removal Act, five Indian tribes were relocated in a decade-long event whose name speaks for itself: it is known as "The Trail of Tears." To understand the meaning of this event, we must reexamine the attitudes of white Americans, particularly those who lived on the western frontier. Their views are nowhere better stated than in these words of President Andrew Jackson, who defending his policy of Indian removal in December 1830 asked the following question: "What good man would prefer a country covered with forests and ranged by a few thousand savages to our extensive Republic, studded with cities, towns and prosperous farms, embellished with all the improvements which art can devise or industry execute, occupied by more than 12,000,000 happy people, and filled with all the blessings of liberty, civilization and religion" (James D. Richardson, ed. *A Compilation of the Messages and Papers of the Presidents, 20* vols. (New York, 1897-1917) 3 (December 6, 1830):1084-85), cited in Horsman [41, p. 202].

In the early years after the Revolutionary War some citizens, among them George Washington, felt that the best way to solve "the Indian problem" was to "civilize the Native Americans" through conversion to Christianity, learning English, and adopting individual ownership of land. In the southeastern United States, the Cherokees, Chickasaws, Choctaws, Creeks and Seminoles did just that, and became known as the "Five Civilized Tribes." Their land was located in parts of Georgia, Alabama, North Carolina, Florida and Tennessee. But as large numbers of white settlers began to move into these states, they wanted the land inhabited by these tribes because they saw that they could become rich by growing cotton. Cotton in particular was a growth industry whose potential profitability was being greatly enhanced by the cotton gin and the institution of chattel slavery. The fact that these five tribes had peacefully given up their own cultural practices in order to do their best to live according to "the American way" did not protect them from suffering the catastrophic consequences of Jackson's Indian Removal Act.

In the winter of 1831, the Choctaw became the first nation to be expelled from its land; they were soon followed by the Creeks in 1836. Forced marches resulted in the deaths of thousands as they were sent -- under strictest coercion -- to Indian Territory in the west. The Cherokees, despite their successful appeals to the United States Supreme Court, were the last to be removed. In May and June of 1838, fifteen thousand Cherokees were rounded up by United

States troops that had been sent by President Martin Van Buren under the command of General Winfield Scott. The Cherokee "were first crowded into thirty-one stockades and then concentrated into eleven camps in Alabama and Tennessee. During their three months in these concentration camps, even before being sent onto the now-infamous Trail of Tears, thousands of Cherokees died" (Bruchac [18], p. 37). After this initial ordeal, the Cherokees were then forcibly marched more than 1,200 miles westward to Indian Territory. Whooping cough, typhus, dysentery cholera and starvation took a heavy toll, and historians estimate that more than 5000 Cherokee died on the Trail of Tears. It should be noted that Indian removal took place in the Northern states as well. In Illinois and Wisconsin, for example, the bloody Black Hawk War in 1832 (Limerick [42], p. 35) opened to white settlement millions of acres of land that had belonged to the Sauk, Fox and other native nations.

Historian Thomas Hietala [31] has called the 1840's "the late Jacksonian period." Building on the land acquired through the Indian Removal Act, the United States experienced an even larger growth and expansion in its territorial possessions. During the late Jacksonian period the administrations of John Tyler and James Polk succeeded in acquiring eight hundred million acres of land for the United States. In 1843 President John Tyler secured the Annexation of Texas in a secretly negotiated treaty signed in April 1844, though it was not supported by members of congress who feared the admission of another slave holding state to the Union [43]. Texas was annexed in 1845; Oregon south of the forty-ninth parallel was acquired in 1846. The Mexican Cession of 1848 added to the continental United States territory equal to the present day states of California, Nevada, Utah, significant portions of Arizona and New Mexico, as well as parts of Colorado and Wyoming. In addition, the U.S. government acquired millions of acres of land in the Great Lakes region surrendered by Native Americans who moved westward to the Great Plains. Hietala [31] points out that "never before had the nation obtained so much territory so quickly…in fewer than a thousand days Tyler, Polk and their supporters pushed the boundaries of the United States to the Rio Grande, the Pacific and the forty-ninth parallel." Emboldened by these successes, initiatives were also pursued to obtain territory and commercial advantages beyond the continent in such places as Hawaii, China, Cuba and Yucatan.

A Note on 19th Century "Scientific Racism"

The joyful and triumphant American westward expansion had its dark, even ominous side. In the late 18th and early 19th century the rhetoric of American politics was still framed by the liberating struggle of the War of Independence and the founding of the Republic. Benjamin Franklin, who used Iroquois symbols both in his language and his design for early American currency, wrote in 1770, "Happiness is more generally and equally diffused among Savages than in civilized societies. No European who has tasted savage life can afterwards bear to live in our societies" (Labaree [44] cited in Hartman 2004 [38], p. 28*)*. Speeches in the early nineteenth century did not have the "rampant racialism" that permeates the midcentury debates. However as the mid-century was approached, American progress and destiny were increasingly explained and understood through arrogant, self-serving concepts of race.

As historian Reginald Horsman [41] states at the outset of his compelling study *Race and Manifest Destiny* (1981): "By 1850 American expansion was viewed in the United States less as a victory for the principles of free democratic republicanism than as evidence of the innate superiority of the American Anglo-Saxon branch of the Caucasian race." His historical analysis traces the misuse of the term "Anglo-Saxon," showing how it was borrowed by Americans from its European origins and transformed by them into one of the key terms used in a potent racial mythology that came to be basis of what most educated Americans in the 19th century viewed as "the scientific study of race." Examined from our perspective today, the doctrine of race that came to dominate in mid and late 19th century America is so extreme that Horsman [41, p. 306] insists on using the terms *"racialism*" and *racialist,"* arguing that our current terms "*racism"* and *"racist,"* can cause confusion when applied to the mid and late 19th century. Furthermore he argues that the obsession with the race concept and the forms in which it was understood during that period are much better understood through the use this alternative terminology.

From about the 1840's through at least to the end of the 19th century, what was called "the scientific study of race" complacently assured white Americans (and anyone else who was listening) that there were superior and inferior races, and with that distinction in hand, one could easily explain all the complexities of history, economics and culture. As an example of this paradigm of *racialism,* consider these words from Josiah C. Nott's *Two*

Lectures (1844): "History and observation both teach that...the Mongol, the Malay, the Indian and the Negro, are now and have been in all ages and places inferior to the Caucasian" (cited in Horsman [41], p. 116). Nott, incidentally, was no fringe figure, but a respected Alabama physician, and "the leading American ethnologist in the 1840's and 1850's" (Muhammad [10], p. 22). In a later extensive work entitled *Types of Mankind* (1854, coauthored with Egyptologist George Glidden), it was strongly argued that Native Americans had no future: "It is as clear as the sun at noon-day, that in a few generations more the last of these Red men will be numbered with the dead...To one who has lived among American Indians, it is vain to talk of civilizing them. You might as well attempt to change the nature of the buffalo."

By 1850 - and for decades to come - the inherent inequality of races was accepted in America without question as a scientific fact. It is true that divergent theories, both religious and non-religious, competed with each other for public acceptance. For instance, the polygenesists argued that it was not possible that the superior Anglo-Saxon race could have been created at the same time as the other inferior races. The Philadelphia physician Samuel George Morton, who possessed the world's largest scientific collection of skulls, published his *Crania Americana* in 1839; "by the end of the 1840's he contended that there had been various Creations in different parts of the world" (Horsman [41], p. 125). When the eminent Swiss-born, European trained biologist Louis Agassiz emigrated to America in 1846, he visited Morton in Philadelphia, corresponded with both Nott and Morton, and four years later allowed them to use his great international prestige when he endorsed their idea of separate creations for different human races. However, some scholars (joined by many clergy and laymen) disagreed vehemently: "the most general disagreement with Nott and Morton was not that they had divided the world into superior and inferior races, but that in adopting polygenesis as the original reason for racial differences, they had challenged the Mosaic account of Creation" (Horsman [41], p. 133).

The American Indians were the main subject of Morton's work with skulls, and his perhaps unsurprising conclusion was that "the intellectual faculties of this great family appear to be of a decidedly inferior cast when compared with those of the Caucasian or Mongolian races" (Horsman [41], p. 127). Based on his measurement of skull sizes of 144 Indian skulls in his collection, Morton (1839: 81) concluded (cited in Gould [45], p. 57) "The structure of his mind appears to be different from that of the white man, nor can the two harmonize in the social relations except on the most limited scale...for the most part [they] are incapable of reasoning on abstract subjects." Unfortunately, Morton not only failed to correct for differences in sex or body size among his skull sample; he did not even recognize the relationship, though it was right there in his data (Gould [45], p. 62).

At first glance the assumptions made by Morton and others assumptions about the direct relationship between measured size of cranial capacity and the level of intelligence seemed to bear fruit. The great French naturalist and zoologist Georges Cuvier sized in at 1,830 grams; the brain of the respected Russian novelist Ivan Turgenev exceeded 2,000 grams. However, later studies by Broca and others turned up the confusing finding that small-brained men of eminence could be found among Caucasians. Thus, as Gould [45, p. 92] unkindly points out, "Walt Whitman managed to hear America singing with only 1,281 grams," and the tiny brain of the French poet, journalist and man of letters Anatole France was measured at just 1,017 grams (for a detailed critical discussion of Morton's procedures, reasoning and results (Gould [45], pp. 50-70).

Phrenology, which claimed to be able to make inferences about cerebral organization based on measurements of different parts of the head, was very popular in the 1840's and 1850's: practitioners traveled, carried charts, casts of heads and handbooks, prepared to give a lecture, and perhaps even to read a head (for a fee). The Fowler brothers, two leading American phrenologists, did not hold out much hope for the Native Americans: "Their small amount of brain in the coronal region of the head, when compared with their immense development of the animal passions and selfish feelings, would bring them chiefly under the dominion of the animal nature of man, and render then little susceptible of becoming civilized, humanized, and educated" (cited in Horsman [41], p. 145). Civilization among the Cherokee had been possible, according to the Fowlers, because the animal portion of members of this tribe was smaller, and the human reasoning portion larger.

The same sort of "save the hypothesis" thinking can be seen in Morton's 1844 work – the *Crania Egyptica* – in which he attempted to show that the Egyptians, who had produced a great civilization, were unquestionably members of the Caucasian and not the Negro race (as some others had suggested). The previously mentioned Dr. Nott was a great admirer of Morton's work, as can be seen in this summary written in 1849: "Dr. S.G. Morton, by a long series of well-conceived experiments, has established the fact, that the capacity of the crania of the Mongol, Indian and Negro, and all dark-skinned races, is smaller than that of the pure white man. And this deficiency seems to be especially well-marked in those parts of the brain which have been assigned to the moral and intellectual faculties" (cited in Horsman [41], p. 133).

Between 1815 and 1850 the transformation in scientific ideas about race was striking; perhaps as a consequence, the new ideas we have attempted to sketch became institutionalized in the periodicals and public debates. The few brave enough to publicly question this prevailing consensus, such as the Presbyterian minister Samuel Stanhope Smith, suffered unrelenting attacks on their views from representatives of the reigning orthodoxy. The superiority of the Anglo-Saxons and the Caucasian race was innate and anyone who questioned this must be a fool. The Anglo-Saxons and Caucasians alone were responsible for civilization in the world. The inferior races would be overwhelmed, would become servants and laborers, or would just disappear. These ideas were found not just in magazines and periodicals, but formed an essential part of the accepted truth found in America's schoolbooks. And of course, these theories were aimed not just at Native Americans, but at all other groups with whom expanding America was or might be in conflict. Thus in 1847, in a speech placed in the Congressional Record, one congressman from Ohio condemned the Mexicans as a "sad compound of Spanish, English, Indian and negro bloods…resulting, it is said, in the production of a slothful, indolent, ignorant race of beings" (cited in Horsman [41], p. 240, p. 347).

As Southerners became increasingly sensitive on the subject of slavery in the 1830's, the tone of their arguments often became strident and defensive, showing a hyperbole that could verge on outright hysteria. Take, for example, these words from William H. Roane, a speaker in the debate over slavery that took place in the Virginia legislature in 1832: "I no more believe that the flat-nosed, woolly-headed black native of the deserts of Africa, is equal to the straight-haired white man of Europe, than I believe the stupid, scentless greyhound is equal to the noble generous dog of Newfoundland" (cited in Robert [46], p. 81). Southern lawyer William Drayton wrote in an anti-abolitionist pamphlet in 1836 that "personal observation must convince every candid man, that the negro is constitutionally indolent, voluptuous, and prone to vice, that his mind is heavy, dull and unambitious; that the doom that has made the African in all ages and countries, a slave – is the natural consequence of the inferiority of his character" (cited in Muhammad [10], p. 21). Racial stigmatization was an equal opportunity employer in the sense that comparable expressions and evaluations were made of all groups and cultures who stood in the way of American expansion and Manifest Destiny.

From our contemporary perspective, the confusion of race with culture, the malignant projections based on narcissistic, self-serving points of view, and the ill will that gives rise to distortions based on close-minded ignorance, are painfully obvious in the above remarks. Viewed from a human development perspective, they perhaps embody at least part of the harmful and life-denying orientation the psychoanalyst Erich Fromm [47] has called "the syndrome of decay." Yet even from the point of view of biological science itself, every single element of what in the 19th century passed for "the scientific study of race" has been discredited. As the preeminent Stanford biologist and taxonomist Paul Ehrlich and developmental psychologist Shirley Feldman [48, p. 38) have written: "*Biologically*, there are no races of *Homo sapiens*." Any proposed racial division of the human species "depends on which characteristic or characteristics are chosen by the classifier." Racial classifications have primarily been on the basis of skin color. But there are literally thousands of traits that could have been chosen instead of skin color: nose shape, height, hair form, frequency of ABO blood group genes, single nucleotide polymorphisms, ability to wiggle the ears, immune system variants, body hair, and countless others.

If race represented a true taxonomic unit, the differences in skin color among human populations would vary concordantly with a multitude other possible traits; in fact, all patterns of geographic variation among human populations vary discordantly. This means that any system that divides human populations into separate units called "races" winds up with a system that is not a unit in nature, but is only an artifact of the trait or traits that through the classifier's arbitrary choice has been used to generate that system. One interesting exercise would be to forbid the use of skin color terms when applied to human populations, substituting instead the terms "less melanized" and "more melanized" to refer to geographic variation in skin color. A second interesting exercise, following Ehrlich's suggestion, would be to forbid the use of the term "race" in any discussion of human biology, and to require instead the term "discordantly varying population cluster." Use of this alternative terminology might force us to confront some of the taken for granted historical baggage that survives and still accompanies many popular and even academic discussions of human diversity based on the concept of race. It would be difficult to have much pride in finding that one belonged to this or that discordantly varying population cluster (depending on the trait or traits under discussion).

The Concluding Period of the Westward Expansion

During this final and critically important phase of "the hundred years expansion," the process of cultural traumatization set in motion by Jackson's Indian Removal Act of 1830 continued, but took place over a wider geographic area made possible by territorial expansion. The years from the late 1840's through to the time of Turner's classic essay [36] were catastrophic for Native American cultural autonomy. In order to see some of the general features of this period, we will focus on the Oglala and Lakota Sioux of the Black Hills (geologically the oldest mountains in the U.S., reaching across South Dakota and Wyoming). Though the historical experience of each tribe is unique, the following discussion focuses on the Lakota and Oglala. They -- together with other Native-American groups -- experienced some or all of the following six forms of attack on their cultural autonomy.

First, *the right to be on their own land was constantly and aggressively challenged.* The frequent disputes over land ownership, property and access were accompanied by decades of intense military conflict both with government soldiers and armed local militias. In the so-called "Indian Wars" (primarily during the 1850's to 1880's), the Lakota and Oglala were fighting to defend their land and their way of life. They were beyond question fierce and effective fighters. Within a ten year period "the Oglala Lakota led by Red Cloud and Crazy Horse had been responsible for two of the three greatest defeats ever inflicted by the United States Army by Indians" (a third defeat was the protracted, expensive and ultimately inconclusive war with the Seminoles in Florida, see Deloria [49], p. 64, treated in depth in Matthiessen [30], pp. 12-20).

Second, *these significant victories were followed by eventual military defeat, and the signing of peace treaties that were later systematically abrogated.* According to the Treaty of Fort Laramie signed by the U.S. government and by Red Cloud, the Oglala war leader, on November 6, 1868, the Indians were guaranteed, "absolute and undisturbed use of the Great Sioux Reservation…No persons…shall ever be permitted to pass over, settle upon, or reside in territory described in this article, or without consent of the Indians pass through the same…No treaty for the cession of any portion of part of the reservation herein described….shall be of any validity or force…unless executed and signed by at least three-fourths of all the adult male Indians occupying or interested in the same" (cited in Matthiessen [30], p. 12). However in 1876 just eight years after the Indians had secured the Black Hills "in perpetuity" in the Treaty of Fort Laramie, the chief Red Cloud was forced to sign a document that appropriated 22.8 million acres of surrounding territory for the U.S. government "in exchange for subsistence rations for an indefinite period" (Matthiessen [30], p. 13).

Third, *an initial forced resettlement was frequently followed by additional episodes of coercive expulsion from the sacred tribal homeland.* In 1889, President Benjamin Harrison proclaimed in an act that he was dismantling the Great Sioux Reservation that had been established at Fort Laramie in 1868. He created seven new reservations; "the Oglala band, which had been the most hostile, was given the dry and rolling hill country between the Dakota badlands and the Sand Hills of Nebraska, now known as the Pine Ridge Reservation" (Matthiessen [30], p. 19). All remaining Lakota land was given to the new states of North and South Dakota, which had been created one month earlier. Forced resettlement is the loss of plants and bird songs, the disappearance of cultural memories teaching reverence for ancestors. What is lost is not just land, but a deep sense of place.

Fourth, *complete destruction of the traditional Native American communal ownership and guardianship of land was legally institutionalized.* The importance of this fact cannot be overemphasized. The General Allotment Act passed by Congress in 1887 (also called the Dawes Act) required each male Indian to accept an allotment of 160 acres from the reservation, with any land not sold scheduled to wind up in the hands of the government and white settlers. Having no experience of individual land ownership or the surrounding economy, not understanding taxes and mortgages, having frequent difficulties with alcohol, most who were forced into this would lose their land. The Dawes Act "legalized an arrangement in which, during the next half century, the native people all across the country would lose two thirds of their remaining lands by sale and swindle." It was initially aimed at tribes in the western Indian territory, but eventually was used against more than one hundred Indian groups, "destroying not only the unity of Indian nations but the people's tradition of generosity and sharing for the common good" (Matthiessen [30], p. 18).

Fifth, *under President Ulysses Grant's 1868 "Peace Policy," Native American children were forced to enter government-supported Christian boarding schools with a curriculum openly based on deculturation.* While some veterans of the Indian Wars favored outright physical extermination, others, such as Captain Richard H Pratt, had a different motto: "Kill the Indian and save the man" (cited in Smith [13], p. 15). "Transfer the savage-born infant to the surrounding of civilization, and he will grow to possess a civilized language and habit," Pratt wrote, unconscious of the utter condescension in his words. In 1879 he founded the first federally sanctioned boarding school: the Carlisle Industrial Training School, in Pennsylvania: this school provided a model which endured for generations. Within three decades of its opening, nearly 500 such schools extended all the way to California.

Richard Monette, a past President of the Native American Bar Association (who attended such a school) wrote that that they are places "where the sharp rules of immaculate living were instilled through blistered hands and knees on the floor with scouring toothbrushes; where mouths were scrubbed with lye and chlorine solutions for uttering Native words" (Smith [13], p. 14). Native American children were kept as virtual prisoners, separated from their families for most of the year, subjected to widespread sexual and physical abuse, forbidden to speak their native language, not allowed to wear their familiar tribal clothes. Native scholars [12] describe the destruction of their culture as a "soul wound," from which Native Americans have not healed.

Sixth, *the destruction of the vast bison herds accomplished by 1883 represented the final blow to their cultural autonomy*. There were thirty to forty million bison roaming over the North American continent at the beginning of the 19th century; by 1883 the species was driven to near extinction with only a few hundred remaining [35]. Buffalo were not only an indispensable element in their material culture (including diet, clothes, and housing) but were considered magical by the Plains Indians. In the cosmology of many tribes, including Cheyenne, Kiowa, and Arapahoe, the mysterious buffalos were seen as "spiritual" beings that often vanished into the underworld during the winter months and then in the spring suddenly reappeared. The cosmos in which these cyclical journeys, deaths and resurrections took place was not the lifeless western universe of mechanical space-time, but a vividly alive alternate world in which living spirits from all directions could bless -- or remove their blessings from -- the activities and goals of the people. Knowledge of the spirit world was essential to perform the rites whose purpose it was to purify the collectivity of the tribe. In the case of many traditional Plains Indians cultures, the buffalo and its artifacts play a role at or close to the center of their most important rituals and ceremonies, those having goals of individual and collective purification, such as the Sun Dance.

The destruction of the bison herds was a deliberate policy of the U.S. government, crafted in response to the frequent failures of the U.S. cavalry in its military encounters with Native Americans. As U.S. Secretary of the Interior Columbus Delano wrote in 1873, "I would not seriously regret the total disappearance of the buffalo from our western plains, in its effect upon the Indians. I would regard it rather as a means of hastening their sense of dependence upon the products of the soil and their own labors" (cited in Geist [50], p. 85). General William Sherman, commander of the Army in the Sioux territory, wrote General Philip Sheridan, suggesting that all the sportsmen of England and America be invited by the military to participate in a "Great Buffalo Hunt"; many came to take advantage of this invitation. It was a collaborative effort of professional hide hunters, sportsmen from the U.S. and Europe, and soldiers from the U.S. Army that brought the bison to near extinction. The railroad companies encouraged shooting of bison from trains. Sometimes they shot the animals for sport, and these massacres left thousands of carcasses to rot on the plains, a sight that enraged the Native Americans. "Have the white men become children?" asked the Kiowa chief Satanta at Medicine Lodge in 1867, "that they should kill meat and not eat?" (Callaway [51], p. 123).

In 1878 an Episcopal church was set up at the Pine Ridge Reservation. In 1881, the sacred Sun Dance was forbidden on all Sioux reservations; the suppression of other religious ceremonies came soon after. In Matthiessen's [30, p. 17) words "At the Indian agencies, the condition of the Lakota people declined rapidly...the proud horsemen who, less than twenty years before, had decreed their terms to the chastened white men at Fort Laramie were already sedentary, half-starved dependents of the U.S. government, despised by those *wasicu* ['the greedy one,' literally, 'he who takes the fat'] for whom they had no respect." It would be the task of future generations to revitalize, renew and rebuild.

In 1888, a Paiute holy man called Wovoka had an apocalyptic vision during a total eclipse of the sun. According to his vision, the Indians would soon receive a messiah, and the world would be set free of the white American conquest. All Indians now dead would return to reinhabit the earth "driving back with them as they return immense herds of buffalo and elegant wild horses to have for the catching...the Great Spirit promises that the white man will be unable to make gun-powder in the future and all attempts at such will be a failure and that the gunpowder now on hand will be useless as against Indians, as it will not throw a bullet with sufficient force to pass through the skin of an Indian" (lines in a letter from Major James McLaughlin, U.S. Indian agent at Standing Rock to the U.S. commissioner of Indian Affairs in Washington, D.C., October 17, 1890, cited in Geist [50], p. 104). The buffalo would return, and the Indians would be able to reclaim their lands and their way of life. Those who wore a sacred Ghost shirt would be protected from the bullets of the whites.

Wovoka insisted that all Native Americans practice the Ghost Dance, and when it reached the Lakota Sioux on the Pine Ridge Reservation, and was practiced there, it caused consternation among the white Americans. As described by Welch and Stekler [52]," The Ghost Dance was a frenzied affair, with much whirling and shrieking and praying, until people would drop from exhaustion and convulse and have visions." Although it was predominantly a peaceful movement, the fearless resistance it inspired in the dancers caused fear of eventual bloodshed on the part of the agents; the response of the U.S. government was "to outlaw the Ghost Dance and crush the resistance anywhere and everywhere" (Geist [50], p. 105). The tragic events that unfolded culminated in the massacre at Wounded Knee on December 29, 1890. More than 200 Lakota women, children and mostly unarmed men were massacred by the U.S Seventh Cavalry, and the subsequent burial while some were still alive in a mass grave, left in Braveheart's words, "a deep psychic wound" that is still part of the historical traumatization of Native Americans today [53]. What Braveheart and her colleagues [26, p. 3) call the *historical trauma response* refers to "a constellation of features that have been observed in massively traumatized populations, including depressive symptoms, psychic numbing, self-destructive behavior, and identification with the dead, among whom vitality in life is seen as a betrayal of ancestors who suffered so much."

We cannot conclude this section of the paper without returning to the events of the Oklahoma Land Rush in 1889. The amorphously labeled "Indian Territory" was a large area that once encompassed what is modern-day Oklahoma. Throughout most of the 19th century, it was thought to be an ideal place to relocate Native Americans who were being removed from their traditional lands to make it open for white settlement. These relocations began in 1817; by the 1880's, "Indian Territory" had become home to a variety of tribes. Yet during the 1880's improved agricultural and ranching techniques had led white Americans to realize that this Indian Territory land should not be left to the Indians because now it could be valuable for white settlement. President Benjamin Harrison was convinced, and in 1889 he authorized a series of actions that eventually removed most of "Indian Territory" from control by Indians. The first of these, which came to be known as the Oklahoma Land Rush, opened the 1.9 million-acre tract of Indian Territory for settlement precisely at noon on April 22nd. Among the displaced tribes were members of four of the five "Civilized Tribes" we have met earlier, who had been forced to leave their lands under Andrew Jackson's Indian Removal Act in 1830. We meet them again now. The Chickasaw, Choctaw, Cherokee and Creek tribes were among those forced to leave their land yet again, this time to make way for the Oklahoma Land Rush. With this our analysis of historical traumatization and loss of cultural autonomy among Native Americans has come full circle.

"The Shining City on a Hill": Manifest Destiny, the Myth of American Exceptionalism, and the Significance of Native American Renewal

By the late 1880s, Native American cultural autonomy had been broken. In Turner's 1893 paper, the frontier had been "the outer edge of the wave – the meeting point between savagery and civilization" (Turner [36], pp. 32-33). It may not be out of the question to ask exactly who were the savages, and who were the representatives of civilization? However one chooses to answer this question, the closing of the frontier discussed by Turner did not mean the end of either the frontier, or of Manifest Destiny. It is beyond the scope of this paper to treat this in the depth it deserves, yet it must be touched upon because of its relevance for present day American history. While we do not hear much talk today of Manifest Destiny, we cannot escape the phrase which is its lineal and direct descendent: *American exceptionalism.*

What are the origins of American exceptionalism? As early as 1630, Massachusetts Governor John Winthrop requested his people "to Consider that wee shall be as a Citty upon a Hill, the eies of all people are upon us" (cited as old English in MacDougall [54], p. 17). Borrowing a New Testament image from the Sermon on the Mount, written on board the ship *Arbella* bound for the New World, it implied that the colonists "were a new Israel entering a new Promised Land" (MacDougall [54] p. 11). In many conventional texts over time the American national identity is traced back to the Puritans, and to this religiously-toned image of "the city upon a hill." The image has political currency in our own time. It was used in a farewell speech by President-Elect John F. Kennedy to the Massachusetts General Court. In 1976 Ronald Reagan added an adjective to Winthrop's phrase, and it became "The Shining City on a Hill," both an effective political mantra and an image that eloquently speaks to the image of American exceptionalism.

The frontier mythology of national expansion continues and is in our time justified by using the language of American exceptionalism. We Americans are "the indispensable nation," "God's country," the "redeemer nation." MacDougall points out that one part of American exceptionalism means that its power, values and wealth and status – backed by its own sense of its unique virtue -- render the United States exempt from the rules and behavior it makes and enforces on other nations. A second, equally important part of the mythology of exceptionalism goes back to the 19th century experience of the frontier; it was and is the deep, almost religious conviction that in America -- and for America --there can be no limits to growth. Perhaps just as important as what Americans saw was what they did not see. The early 19th century French visitor Alexis de Tocqueville observed that Americans are "insensible to the wonders of inanimate nature, and they may be said not to perceive the mighty forests which surround them until they fall beneath the hatchet" (cited in Bellows [55], p. 112). The blindness observed by de Tocqueville continues today, but we are no longer living in a time where we can live in a culturally-induced slumber supported by the delusional belief in a world without limits. As the contemporary historian and sociologist, Robinson [56] has written: "Our world is burning. We face a global crisis that is unprecedented in terms of its magnitude, and scale of the means of violence. This is a time of great upheavals, momentous changes, and uncertain outcomes fraught with dangers."

Blinded by the frontier heritage and the complacent belief in American exceptionalism, while the world burns our newest frontier wars are being fought elsewhere, in Iraq, Afghanistan, Syria, Libya, Yemen and Somalia. Firm in the belief that these actions are increasing our national security, we wage unaccountable war from the air in these countries with drones that induce fear before they rocket down into villages and towns. At the same time we wage unceasingly a destructive war on the global environment. Leading environmental scientists have identified nine "planetary boundaries" necessary to maintaining an earth conducive to human existence: climate change, ocean acidification, stratospheric ozone depletion, the nitrogen and phosphorous cycles, global freshwater use, change in land use, biodiversity loss, atmospheric aerosol loading and chemical pollution (Robinson [56], p. 230, Foster, Clark and York [57], p. 14). All these boundaries are threatened.

The mythology of American exceptionalism, with its mandatory belief in the inevitable growth and expansion of American power, has the quality of a deeply rooted cultural obsession. Conditioned by the 19th century frontier experience, it sometimes seems as if we – the American people – have not arrived and do not actually live here on this land. In contrast to the pioneer myth of progress, our two centuries old relationship with the land is better described in the troubling phrase used by Kentucky farmer and poet Wendell Berry [58]: "the unsettling of America." As the founder of the Land Institute, Wes Jackson [59, p. 2) wrote in a book dedicated to Berry: "It has never been our goal to become native to this place. It has never seemed necessary even to begin on such a journey." Going one step further, the anthropologist Loren Eiseley [60, p. 53] has compared us to slime moulds: "It came to me in the night, in the midst of a bad dream, that perhaps man, like the blight descending on a fruit, is by nature a parasite, a spore bearer, a world eater."

Other and earlier civilizations have collapsed under the impact of their blindness and inability to foresee approaching environmental catastrophe: Sumer, Easter Island, the Mayan city-states, the Roman Empire [61]. Yet as Robinson [56, p. 230] points out, other civilizations have been in a destructive relationship with nature. What may be unique this time is "not that it [our society] is in fundamental contradiction with nature. Rather it is **the scope and magnitude of this contradiction**, such that human activity now threatens the earth system itself. The current moment is unique in that this time the collapse would be that of global civilization."

It is a profound historical irony that the wisdom of the traditional Native American cultures – for the most part ignored and relegated to the scrap heap of history by mainstream American culture – may prove indispensable for the health, even the survival of mainstream cultures facing "the global crisis of humanity." For during the past five decades, there has been a profound historical movement and transformation among both North American Native Americans and other indigenous people worldwide. Native American teachings on law and democracy (Chief Oren Lyons), food and agriculture (John Mohawk), energy development (Clayton Thomas-Mueller); and environmental and reproductive justice (Katsi Cook) are only a few of the articles in Melissa K. Nelson's *Original Instructions: Indigenous Teachings for a Sustainable Future* [62]. These articles, which give a small sample of contemporary Native American renewal, revitalization and rebuilding, are based on talks given at the Annual Fall Bioneers Conference held in San Rafael, CA. As Bioneers co-founder Kenny Ausubel observes in the Preface to Nelson's edited collection, these talks and others in related volumes "embody some of the most ancient wisdom on earth from the world's 'old-growth cultures'' (see for example Harney [1], Stone and Barlow [63]).

Preceding sections of this paper have described events in 19th century American history whose effects on Native Americans cannot be understood without employing the paradigm of historical traumatization. In his seminal work *The Sociological Imagination*) the sociologist C. Wright Mills [63, p. 7] asked these two questions: "What varieties of men and women now prevail in this society and in this period? And what varieties are coming to prevail?" The Native Americans actively engaged in the tasks of renewing, revitalizing and rebuilding constructive responses to historical traumatization have all become a new variety of men and women. They remain deeply true to their indigenous tribal heritage and its teachings; yet they have also learned widely and wisely from Western knowledge in multiple fields. Having this level of cognitive and emotional integration, they can well be called exemplars of "the bicultural personality."

Thus the Mohawk Indian midwife Katsi Cook was delivering babies to a family who lived within a mile of a Superfund site when she was asked by a mother: "Is it safe to breast feed?" Finding no answers, her research emerged into the first human health study at a Superfund site that brought together health researchers, community members and health care providers. Mohawk women themselves are "co-investigators in the scientific research; we all share power and authorship of scientific publications" (Cook [65], p. 8, see also Cook's contribution to Nelson [62]. Cook, who has been invited to teach and lecture at mainstream U.S. medical schools, has not only brought back the Mohawk midwife tradition, but has been instrumental in empowering women from the Oneida, Onondaga, Tuscarora and Cayuga tribes to have their babies born on their own lands, using their indigenous traditions and culture. Cook [65], p. 18] writes, "The door of life and the door of death are the same door, and when you lose the knowledge of how to be born, you lose the knowledge of how to die."

An early illustration of the paradigm of interconnectedness on which this paper is based can be seen in the following statement by Chief Seattle (1780-1866): "Humankind has not woven the web of life. We are but one thread within it. Whatever we do to the web of life we do to ourselves. All things are bound together. All things connect" (cited in Anastasia [39], p. 145). Our collective task as Americans is to apply this paradigm of interconnectedness to our historical relationships with the rest of the global community, vowing to protect the integrity of the single miraculous blue green planet, with the same vow that is made in traditional Native American cultures: down to the seventh generation. The probable consequences of a failure to do so can be seen in these lines of Shelley's poem Ozymandias [66]:

> I met a traveller from an antique land
> Who said: Two vast and trunkless legs of stone
> Stand in the desert. Near them on the sand,
> Half sunk, a shatter'd visage lies, whose frown
> And wrinkled lip and sneer of cold command
> Tell that its sculptor well those passions read
> Which yet survive, stamp'd on these lifeless things,

The hand that mock'd them and the heart that fed.
And on the pedestal these words appear:
"My name is Ozymandias, king of kings:
Look on my works, ye Mighty, and despair!"
Nothing beside remains: round the decay
Of that colossal wreck, boundless and bare,
The lone and level sands stretch far away.

ACKNOWLEDGMENTS

We thank Iona College, California State University East Bay, National Park Service and Dominican University of California for supporting our work. We express our deep appreciation to Joyce Anastasia for her detailed reading and helpful comments on the pre-publication manuscript.

REFERENCES

[1] Harney, C. The Way It is. Nevada City, CA: Blue Dolphin Publishing (2009).

[2] McGinley, E; Varchevker, A. (Eds.) Enduring Trauma Through The Life Cycle. New York: Karnac, (2013).

[3] Straussner, SL. Trauma Through the Life Cycle: A Review of Current Literature, *Clinical Social Work Journal*, 42(4), 323-335 (2014). DOI: 10.1007/s10615-014-0496.

[4] Levy, BS; Sidel, VW. Adverse health consequences of the Iraq War. *Lancet*, 381 (9870), 949-958 (2013). DOI: 10.1016/S0140-6736(13)60254-8.

[5] Cardozo, BL; Bilukha, OO; Crawford, CA. Mental health, social functioning, and disability in postwar Afghanistan. *JAMA*, 292:575–584 (2004).

[6] Amir, M; Lev-Wiesel, R. Time does not heal all wounds: Quality of life and psychological distress of people who survived the Holocaust as children 55 years later *Journal of Traumatic Stress*, 16(3), 295-299 (2003). DOI: 10.1023/A:1023756326443.

[7] Lifton, RJ. Death in Life: Survivors of Hiroshima. Chapel Hill, NC: University of North Carolina Press (first published 1968) (1991).

[8] Ngo, AD; Taylor, R; Roberts, CL; Nguyen, TV. Association between Agent Orange and birth defects: systematic review and meta-analysis. *International Journal of Epidemiology*, Vol. 35 (5), 1220-1230 (2006).
[9] Gilbert, S. Developmental Biology. 10th Edition, Sunderland, MA: Sinauer, (2013).
[10] Muhammad, KG. The Condemnation of Blackness, Cambridge, MA: Harvard University Press, (2010).
[11] Alexander, M. The New Jim Crow: Mass Incarceration in the Age of Colorblindedness. New York: The New Press, (2012).
[12] Duran, E. Healing the Soul Wound: Counseling With American Indians and Other Native Peoples. New York, NY: Teachers College Press, (2006).
[13] Smith, A. Soul Wound: The legacy of Native American Schools. *Amnesty Now*, Summer, 14-18 (2003).
[14] Brave Heart, MYH. The historical trauma response among natives and its relationship with substance abuse: a Lakota illustration. *J. Psychoactive Drugs*, 35(1), 7–13 (2003).
[15] Brave Heart, MYH. The return to the sacred path: healing the historical trauma and historical unresolved grief response among the Lakota through a psychoeducational group intervention. *Smith Coll Stud Soc Work*, 68(3), 288–305 (1998).
[16] Brave Heart, MYH. *Oyate Ptayela*: rebuilding the Lakota nation through addressing historical trauma among Lakota parents. *J. Hum Behav Soc Environ.*, 2(1–2), 109–126 (1999).
[17] Segal, CM; Stineback, DC. Puritans, Indians and Manifest Destiny. New York: G.P. Putnam's Sons (1977).
[18] Bruchac, J. The Hawk in The Prison. *Parabola.* Summer, 37-42 (2003).
[19] Bateson, G. Steps to An Ecology of Mind. New York: Ballantine, (1972).
[20] Borish, S. The Land of the Living: The Danish Folk High Schools and Denmark's Non-violent Path to Modernization. Nevada City, CA: Blue Dolphin Publication, (1991)
[21] Hansgaard, T; Landboreformerne, I; Danmark, I. det 18. Aarhundrede: problemer og synspunkter (The Land Reforms in Denmark in the 18th century: problems and viewpoints). Copenhagen: Landbohistorisk Selskab, (1981).
[22] Vorvick, L; Zieve. D. "Fetal alcohol syndrome – PubMed Health." National Center for Biotechnology Information. N.p., n.d. Web. 21 Apr. 2011 at http://www.ncbi.nlm.nih.gov/pubmedhealth.

[23] Dorris, M. The Broken Cord. New York: Harper& Rowe, (1989).
[24] French, L. Addictions and Native Americans, Westport, Conn.: Praeger (2000).
[25] Merrill, D; Zieve, D. "Alcoholism and alcohol abuse – PubMed Health." National Center for Biotechnology Information. N.p., n.d. Web. 21 Apr. 2011 at http://www.ncbi.nlm.nih.gov/pubmedhealth/PMH0001940/.
[26] Brave Heart, MYH; Elkins, J; Tayofa, G; Bird, D; Salvador, M. *Wicasa Was'aka*: Restoring the Traditional Strength of American Indian Boys and Men. *Am J Public Health*, 102 (Suppl 2), S177–S183 (2012). DOI: 10.2105/AJPH.2011.300511. (2012).
[27] Castor, ML; Smyser, MS; Taualii, MM; Park, AN; Lawson, SA; Forquera, RA. A nationwide population-based study identifying health disparities between American Indians/Alaska Natives and the general populations living in select urban counties. *Am. J. Public Health*, 96(8), 1478–1484 (2006).
[28] Sindell, PS. (1997). Some Discontinuities in the Enculturation of Mistassini Cree Children. In G. Spindler, ed. Education and Cultural Process: Anthropological Approaches. Third Edition. Prospect Heights, IL: Waveland. (1997).
[29] Curtis, E; Brown, JE. The North American Indians: Photographs by Edward Curtis. New York: Aperture, (1972).
[30] Matthiessen, P. In the Spirit of Crazy Horse. New York: Viking (1983).
[31] Hietala, TR. Manifest design: American Exceptionalism and Empire. Itaca, NY: Cornell University Press, (1985).
[32] Weeks, WE. Building the Continental Empire: American Expansion from the Revolution to the Civil War. Chicago, IL: Ivan R. Dee (1997).
[33] Moser, DD. (editor), Matthew Lyon to Andrew Jackson, 2 June 1814, The Papers of Andrew Jackson, Knoxville: University of Tennessee Press, 4:78-79 (1994).
[34] Isenberg, AC. The Destruction of the Bison. New York: Cambridge University Press, (2000).
[35] Kolipinski, M; Borish, S; Scott, A; Kozlowski, K; Ghosh, S. Bison: Yesterday, Today and Tomorrow. *Natural Areas Journal*, 34 (3), 365-375 (2014).
[36] Turner, FJ. Rereading Frederick Jackson Turner. New York: Henry Holt (1994).
[37] Owsley, FL; Smith, GA. Filibusters and Expansionists: Jeffersonian Manifest Destiny, 1800-1821. Tuscaloosa, AL: University of Alabama Press, (1997).

[38] Hartmann, T. What Would Jefferson Do? A Return to Democracy, New York: Harmony Books, (2004).
[39] Anastasia, J. Extraordinary Leadership during Extraordinary Times. Mill Valley, CA: Lead by Wisdom, (2015).
[40] Williams, R. Savage Anxieties. New York: Palgrave MacMillan, (2012).
[41] Horsman, R. Race and Manifest Destiny. Cambridge, MA: Harvard University Press, (1981).
[42] Limerick, P. Something in the Soil: Legacies and Reckonings in the New West. New York: Norton. (2000)
[43] Crapol, EP. John Tyler: The Accidental President. Chapel Hill, NC: University of North Carolina Press. (2006)
[44] Labaree, L. Papers of Benjamin Franklin, (New Haven: Yale University Press. "*Benjamin Franklin's Marginalia*, 1770," vol. 17 (1965)
[45] Gould, SJ. The Mismeasure of Man. New York: Norton (1981).
[46] Robert, JC. The Road From Monticello: A Study of the Virginia Slavery Debate of 1832. New York: AMS Press (1941).
[47] Fromm, E. The Heart of Man. New York: Harper and Rowe. (1968).
[48] Ehrlich, P; Feldman, S. The Race Bomb: Skin Color, Prejudice and Intelligence. New York: Ballantine. (1977).
[49] Deloria, V. Jr. Behind the Trail of Broken Treaties. Austin: University of Texas Press, (1985).
[50] Geist, V. Buffalo Nation: History and Legend of the North American bison. Voyageur Press, Stillwater, MN. (1996).
[51] Callaway, CG. ed., Our Hearts fell to the Ground: Plains Indians Views of How the West was Lost. Bedford/St. Martins, Boston, MA. (1996).
[52] Welch, J; Stekler, P. Killing Custer: The Battle of Little Bighorn and the Fate of the Plains Indians. New York: Norton. (1994).
[53] Brown, D. Bury My Heart at Wounded Knee. New York: MacMillan (1971).
[54] McDougall, WA. Is America (Still) Exceptional? *The American Interest: Policy, Politics and Culture.*, Vol VIII, No. 4 March/April, 6-15 (2013).
[55] Bellow, S. Reflections on Alexis de Tocqueville. *The American Scholar.* Vol. 84, No. 2 Spring 110-115 (2015).
[56] Robinson, WI. Global Capitalism and the Crisis of Humanity. New York: Cambridge University Press. (2014).
[57] Foster, JB; Clark, B; York, R. The Ecological Rift: Capitalism's War on the Earth. New York: Monthly Review Press, New York. (2010).
[58] Berry, W. The Unsettling of America. San Francisco: Sierra Club Books (1986).

[59] Jackson, W. Becoming Native to This Place. Washington, D.C – Counterpoint, (1996).
[60] Eiseley, L. The Invisible Pyramid: A Naturalist analyses the rocket century. New York: Charles Scribner's Sons, (1970).
[61] Diamond, J. Collapse: How Societies Choose to Fail or Succeed. New York: Penguin, (2005).
[62] Nelson, M. ed., Original Instructions: Indigenous Teachings for a Sustainable Future. Rochester, VT: *Bear & Company*, (2008).
[63] Stone, MK; Barlow, Z. Ecological Literacy: Educating our Children for a Sustainable World. San Francisco: *Sierra Club Books*, (2005).
[64] Mills, CW. (1999). The Sociological Imagination, First published 1959, New York: Oxford University Press, 40th Anniversary Edition, (1999).
[65] Cook, K. Into Our Hands. Yes: *A Journal of Positive Futures*. Spring 2000, 17-19.
[66] Shelley, PB. The Complete Works of Percy Bysshe Shelley, [Original Poetry. (1818, January 11). The Examiner (London), p. 24.], Roger Ingpen & Walter E. Peck eds. New York: Gordian Press, 1965.

In: Native Americans
Editor: Jerome Mendoza
ISBN: 978-1-63483-184-0

Chapter 2

THE PEOPLING OF AMERICA: SIBERIANS, PACIFIC ISLANDERS AND ANCIENT TRANS-ATLANTIC EUROPEANS ACCORDING TO HLA GENETICS AND ANTHROPOLOGY

***Antonio Arnaiz-Villena*[1,*], *Ester Muñiz*[1], *Jose Palacio-Gruber and Jorge Martinez-Laso*[2]**
[1]Department of Immunology, University Complutense, School of Medicine, Madrid, Regional Blood Center, Madrid, Spain
[2]Department of Immunogenetics. Instituto de Salud Carlos III, Madrid, Spain

ABSTRACT

America peopling has recently been explained based only on genetic data. While different First America inhabitants' ethnic groups, Amerindians, Na-Dene speakers, Aleuts and Eskimo there exist, there is no genetic, cultural or anthropological homogeneity within these groups. We have particularly addressed the relatedness of First America

* Correspondence Adress: Antonio Arnaiz-Villena. Departamento de Inmunología, Facultad de Medicina, Universidad Complutense, Pabellón 5, planta 4. Avda. Complutense s/n, 28040 Madrid, Spain. Phone +34 91 3017354, Fax +34 91 3017229, e-mail: arnaizantonio @gmail.com.

Inhabitants with Pacific Islanders by using autosomal genetic markers: the HLA alleles. HLA is the most polymorphic human genetic system and this is most useful for comparing population relatedness. Ethnic groups of Pacific Islanders, First America Inhabitants and other World Populations have been used. A genealogic study and also a frequency comparison studies by using HLA alleles and haplotypes have been carried out. Our conclusions are: 1-Aleuts seem to be a genetic and linguistic separate group which may be related to northern European Lapps, both of them originated in southern Siberia Baikal Lake area. 2- First America Inhabitants, including all analyzed Amerindians, Na-Dene speakers and Eskimo have had genetic flow with Pacific Islanders: the latter share autosomal HLA alleles and haplotypes with First America Inhabitants. This could have been bidirectional. 3- Particularly, Easter Islanders show a probable cultural and genetic exchange with Titikaka Lake Aymaras and a contact possibility according to Kon-tiki Lima/Polynesia Pacific expedition. This civilization also shares significant traits with European Iberian megalithic builders. 4- Mesoamericans may be grouped together because they bear more ancient Olmec culture traits and our HLA results. 5- Genetics is not able by itself to uncover in space and time America peopling and First American Inhabitants relatedness with Pacific Islanders and ancient Solutrean Europeans. 6- Megalithic and genetic data found in Azores Islands shows that a pre-Celtic/Iberian culture may have also reached these Islands (and may be America) in Iron/Bronze Neolithic periods. Thus, there are both genetic and cultural solid evidence for Trans-Atlantic contact between Europe and America in ancient times.

Keywords: Afro-Americans, Aleuts, Ancient Europeans, Amerindians, Azores, Bantu-Spanish, Easter Island, Eskimo, Celts, HLA, Haplotypes, Iberians, Kon Tiki, Lapps, Megaliths, Na-Dene, Olmecs, Polynesians, Pacific Islanders, Palaeolithic Writing, Rapa Nui, Solutreans, Siberians

This work partial results were given at a lecture by Antonio Arnaiz-Villena at The Roy al Society (London)- June 9th 2011 "Origin of Amerindians and their genetic relatedness with Asian and Pacific Islanders" (https://rovalsociety.org/events/2011/plagues-pathogens-selection/)- Session 3.

INTRODUCTION

Groups, Genetics and Migration

The First Amerindian Natives are postulated to have come from Asia through the Bering land bridge between 30,000-12,000 years before the present (BP). These conclusions have been based on cultural, morphological and genetic similarities between American and Asian populations. Both Siberia and Mongolia (Kolman et al., 1996; Merriwether et al., 1996) have been put forward as the most likely places of origin in Asia. (Figure 1)

Greenberg first postulated the triple migration theory for explaining the peopling of the Americas (Greenberg et al., 1986): Amerindians (most North and South American Indians; 12,000 years BP), Na-Dene (Athabaskans, Navajo, Apache; 8,000 years BP) and Eskimo-Aleuts (6,000 years BP). Research carried out before the widespread use of Y Chromosome (Y Chr) and other nuclear DNA markers including mtDNA (Wallace & Torroni, 1992) for the study of populations (Cavalli-Sforza et al., 1994; Parham & Ohta, 1996) supported the three-wave model. However, other mtDNA studies have not (Horai et al., 1993; Torroni et al., 1993); other authors postulate only one wave coming from Mongolia / North China as giving rise to the First Native American ancestors (Kolman et al., 1996; Merriwether et al., 1996). The study of Y Chromosome DNA markers seemed to suggest the existence of a single major paternal haplotype in both North and South American Native populations (Karafet et al., 1997; Santos et al., 1996). However, other studies on Y Chromosome show that more than one paternal founder haplotypes arrived in America during different migrations (Karafet et al., 1999), probably from Siberia (Santos et al., 1999).

Alu-insertion investigations have also been carried out to ascertain the origin of First Americans (Novick et al., 1998). The results are not concordant with the multiple-wave migration hypothesis; a surprisingly short genetic distance between Chinese and Native Americans was found and explained by a recent gene flow from Asia (Novick et al., 1998).

More recently, new mtDNA analysis has suggested that all mtDNA lineages must have been isolated in Asia before entering the New World by at least 7-15 thousand years. It is even suggested that this place must have been Beringia (Mulligan et al., 2008). Also, a dispersal of Amerindians coming from Asia has been put forward through Coastal Pacific line (Goebel et al., 2008) based on all available archaeological, anthropological, mtDNA and genetic data.

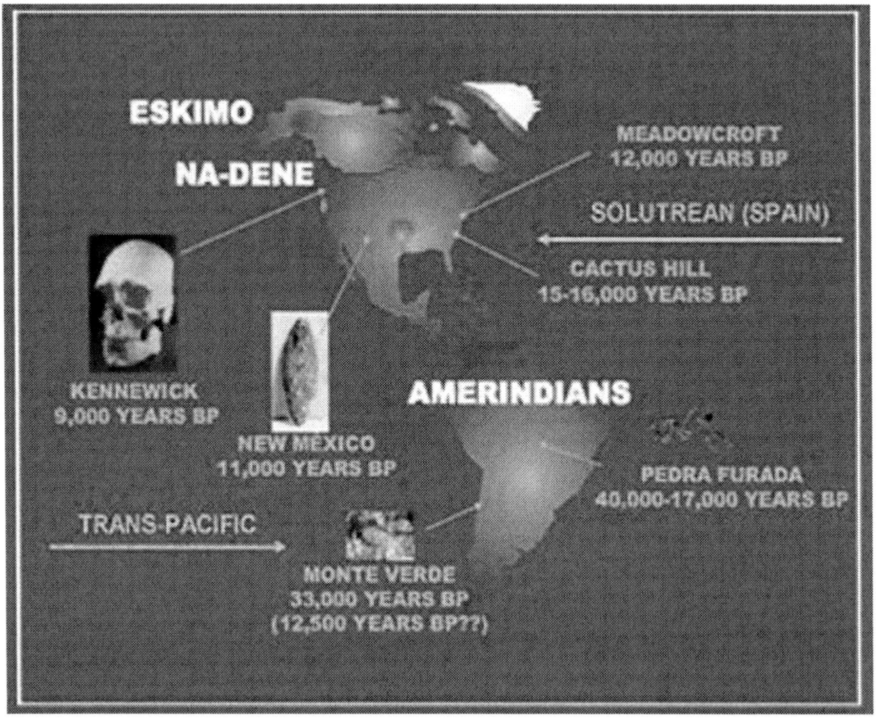

Figure 1. America peopling. *It depicts the most popular theory of peopling of this continent from Asia though Bering Strait (Greenberg et al., 1986). Green: Amerindians (30,000-12,000 years BP); Red: Na-Dene (8,000 years BP), Athabaskans in Canada, Californian Indian isolates and Navajo and Apache from Southern United States; Grey: Eskimo (6,000 years BP). Aleuts from Aleutian Islands in Bering Strait are separate from Eskimo in linguistic and other anthropological parameters and were present in the Islands before Eskimos reached North America; in addition, Aleut HLA profile is different from Eskimo profile(Moscoso et al., 2008). Other theories of peopling of Americas (Yellow arrows): Trans-Pacific (from Australia-Pacific Islands (Cerna et al., 1993)), and from Iberian Peninsula Solutrean people (Bruges-Armas et al., 1999; Holden et al. 1999). Archeological relevant findings are also represented (Holden et al. 1999). Kennewick man from Washington State, USA; Meadow croft (Pennsylvania, USA); Cactus Hill (Virginia, USA); Pedra Furada (Brasil); Monte Verde (Chile).*

In this regard, HLA data may be more informative than Y Chr and mtDNA (Uinuk Ool et al., 2002) because maternal and paternal lineages and both frequencies (i.e.: genetic distances, dendrograms and correspondence analyses) and genealogies (specific HLA alleles and haplotypes) may be studied for comparing populations. The best evidence that HLA is a good genetic marker for studying population relatedness is that it usually correlates with geography.

Paleolithic Age Worldwide Rock Symbols (Figure 2)

In 2010, Genevieve Von Petzinger published a work showing that the same rock symbols are present in all parts of the World from North and South America to Spain, China and Australia. These findings considerably alter many populations and peopling theories around the World, including America peopling. Until this discovery is explained, all postulates about First American Inhabitants must be carefully taken. Implications are that ancient people was moving around World earlier than thought in Palaeolithic times, some of them around 75,000 years ago (Von Petzinger, 2009).

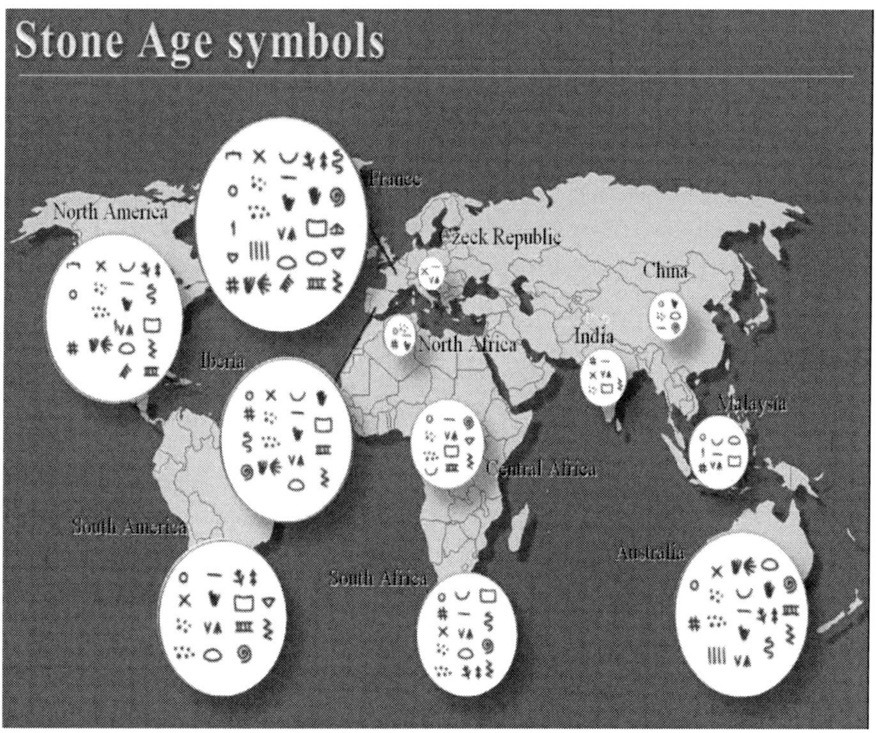

Figure 2. Attributed writing common to all world in Palaeolithic time. *Some of them ancient as 75,000 years ago (Von Petzinger, 2009).*

Trans-Pacific Routes: Easter Island (Rapa Nui in Easter Island Original Language) (Figure 1)

Calculations done by using paternal (Y Chr) or maternal (mtDNA) lineages may be biased when population displacements are concerned, because these movements are carried out by both males and females. In contrast autosomal HLA gene analyses do take into account both sexes (Arnaiz-Villena et al., 2000). This may have happened in analyses of theoretical Amerindians displacement from Asia to the Americas. In addition, other authors (Uinuk-Ool et al., 2002) using nuclear histocompatibility (HLA) markers do not regard as important and possible to establish the number and timing of migration waves. The important issue is whether immigrants (Amerindians) were already differentiated (in Asia) into such ethnic groups whose descendants are still to be found in Asia. If they were differentiated then the question of how and when they crossed the Bering Land Bridge is a secondary one (Uinuk-Ool et al., 2002).

Also, a Trans-Pacific route of American peopling from Asia or Polynesia has been suggested because HTLV-1 virus strains shared identical sequences in Japan and in the northern coast of South America (Leon-S et al., 1996) and some HLA alleles may have been introduced by the same Trans-Pacific route (Arnaiz-Villena et al., 2009; Cerna et al., 1993). In the same way, "quasi-specific" Amerindian HLA alleles, like A*02:12 or B*39:05 (Arnaiz-Villena et al., 2005, Arnaiz-Villena, 2014), have been found in several unrelated individuals of Easter Island, which suggests an early contact between Easter Polynesians Islanders and Amerindians (Lie et al., 2007). Recent genetic studies have identified Polynesian mtDNA haplogroups in remains (skulls) of Botocudo Amerindians from Brazil (Amerindian group extinct by the end of 19th century) (Gonçalves et al., 2013). Other signs may indicate a communication between these groups of populations, like the presence of South American sweet potato in earlier Pacific sites (Lawler, 2010) or the finding of chicken remains of Polynesian type in El Arenal (Chile) dated by radiocarbon back to 1300-1400 AD (Storey et al., 2007; Storey et al., 2008). Furthermore, skeletal remains of pre-Columbian individuals with Polynesian ancestry and several Mapuche artefacts which are similar to Polynesian ones at Mocha Island (Chile) have been recently reported (Lawler, 2010; Matisoo-Smith & Ramirez, 2010). All these facts provide evidences for this Trans-Pacific route that could have occurred in both ways at different times. It is classical to mention the Kon Tiki expedition that showed that Amerindians could reach Polynesia by boat and that Easter Island culture and Titikaka Lake-Tiwanaku culture may be related and also genetic relationship may occur

(Arnaiz-Villena et al., 2014). In spite of these efforts to relate genetically both cultures, a more simple and feasible explanation to common Amerindian and Easter Island genetic traits (Moreno-Mayar et al., 2014). Peruvian slave traders are documented to have often visited Easter Island in 1862 AD; they took slaves to Peru and most likely admixed with Easter Island Inhabitants. This has already been put forward (Zink, 1980).

Atlantic European Input before Columbus

1. Atlantic Solutrean Input in America (Figure 1)

Stanford and Bradley have widely documented archaeological findings in several USA sites (particularly Clovis, New Mexico; Cactus Hill, Virginia) that show existence of ancient Solutrean Europe-America cultural exchanges around 21000 years ago. Also a mitochondrial DNA lineage is also shared between First American Inhabitants (from early pre-Columbian remains in an eastern North-America) and Europeans. This lineage does not occur in Asia in prehistoric times (Reidla et al., 2003).

2. Basque/American contact before Columbus

a. St Lawrence River/Newfoundland

Basque fishermen reached St Lawrence River Mouth and Newfoundland since uncertain times. Although the first documented Basque ballen hunt news by Basques came from XIth AD. Algonquin, Mic-Mac, Mountagnais Amerindians from these areas early established a Basque-Amerindian pidgin for addressing to Europeans. Lescarbot in his "*History of New France*" (1612) quotes this fact widely. Basques are documented to have been established themselves for trading factories and fisheries around St Lawrence River Mouth and Newfoundland early in XVIth century AD. First settlement is recorded in 1517, but earlier contacts are not discarded. Also an Icelandic/Basque pidgin was also established (Bakker et al., 1991).

The facts are that a number of Basque settlements have been found around St Lawrence River Mouth, Labrador Peninsula and Newfoundland. Basque toponimics are found in Newfoundland and all over Quebec provinces (Canada).

French explorers met Algonquinian Indians at St Lawrence River Mouth who spoke to him with Basque words and constructions. This showed that Basques have been trading with them for a long time (Bruges-Armas, 1999).

b. Andean/Basque (Iberian-Tartessian) language similarities

Ancient Basque is probably one of the most primitive languages in European/Mediterranean area, which apparently was much more widespread, forming a part of the NaDene-Caucasian group of languages (Ruhlen, 1991). We have named Mediterranean languages as Usko-Mediterranean languages (Arnaiz-Villena et al., 2001) (Figure x) and Untermann has named a probably similar concept as "languages with Mediterranean Stratum" (Untermann, 2006, pp. 197). American Na-Dene, Athabaskans, Navajo and Apache languages in North America belong to this type of languages. In addition, results are shown in Table 1 of Andean (Aymara/Quechua) words that are similar in phonology and semantics to Basque.

Table 1. Iberian-Andean similar language. http://chopo.pntic.mec.es/biolmol/publicaciones/Usko.pdf

Aymara/ /Quechua	English	Basque (Iberian-Tartessian)	English
Titi	(1st part of Titikaka)	Tita*	Sin, freckle *
Kaka	(2nd part of Titikaka)	Kaka	Human remains, deads *
Viracocha	Deity who created everything	Bira	Two by two
		Kotxa	Fertile
Mitimaes	Almost all people living in Cuzco at the Spanish Conquest timing (according to Cieza de León)	Miti	Servant
		Maes	Corn (Corn harvesters)
Curaca	Religious-administrative authority	Kur	Adorer
		Aka	Dead
Ilapa	Thunder's God	Il	Dead
		Apa	Father (Dead's Father)
Ceque	Earth paths related with ritual sacrifices	S(Z)equiri	Crowd following deads to be buried
Huaka	Sacred place	Gu(ne)aka	Our deads (cemetery?)
Ekeko	Benefical man wondering on "Altiplano"	Ekei	Much grown up person
		Ko	Function
Panakas	Dead King's family	Banako	Special, choosen people.
Urko	Viracocha's son	Urk(h)o	Close relative
Kipu	Andean rope with knots for writing and calculations	Kipustun	Knot's lace.

These terms have been analyzed by using a methodology described in http://chopo.pntic.mec.es/~biolmol/publicaciones/Usko.pdf supported on phonology and semantics of the same words, which is also pointed to be fruitful by Untermann. (2006, pp. 198), (Arnaiz-Villena & Rey, 2012). http://euskararenjatorria.net/? page_id =13028&lang=es. (First International Basque-Iberism Congress, Gernika, 2014)

3. The problem of Azores findings

We have already quoted that both genetic (Bruges-Armas et al., 1999) and archaeological (Holden, 1999; Stanford & Bradley, 2012) evidence suggests that a two-way Trans-Atlantic traffic occurred before Columbus discovered America. Some coins from Corvo Island (Azores) from Carthago and Cyrene have been recorded by JF Podolyn in XVIIIth century AD (Patricia and Bikal, 1990). Also, dozens of pre-Christian megalithic hypogea and tombs have been found in Azores Islands: Terceira, Corvo and Flores (Rodriguez, 2015); they are from Mediterranean origin and Bronze or Iron Age. They should be included in the Atlantic/Mediterranean Megalithic culture. In Atlantic Europe, megaliths are found from Scotland to southern Spain. These Azores findings are in fact megalithic corridor tombs, frequently showing wall rock engravings similar to Spanish or Irish ones (Arnaiz-Villena et al., 2013). In conclusion Azores culture should be included within the Celtic/Iberian Atlantic area culture (http://basques-iberians.blogspot.com.es/2013/11/son-iberos-los-celtas.html)

In the last 15 years, we have studied the North, Meso and South American Amerindians' HLA allele frequencies and have compared them with those of other North American First Inhabitants and Asians, particularly with Central and South East Asia, and Pacific populations (Arnaiz-Villena et al., 2000). HLA genes have been analyzed for the following Amerindian ethnic groups: Mayans (Gomez-Casado et al., 2003), Mixe, Mixtecans, Zapotecans (Petzl-Erler et al., 1997), Lakota Sioux (Leffell et al., 2004), Mazatecans (Arnaiz-Villena et al., 2000), Lamas (Moscoso et al., 2006), Quechuas (Martinez-Laso et al., 2006), Aymaras (Arnaiz-Villena et al., 2005), Uros (Arnaiz-Villena et al., 2009), Tarahumaras (Garcia-Ortiz et al., 2006), Mapuches (Rey et al., 2013), Toba Pilaga, Mataco Wichi, and Eastern Toba (Cerna et al., 1993).

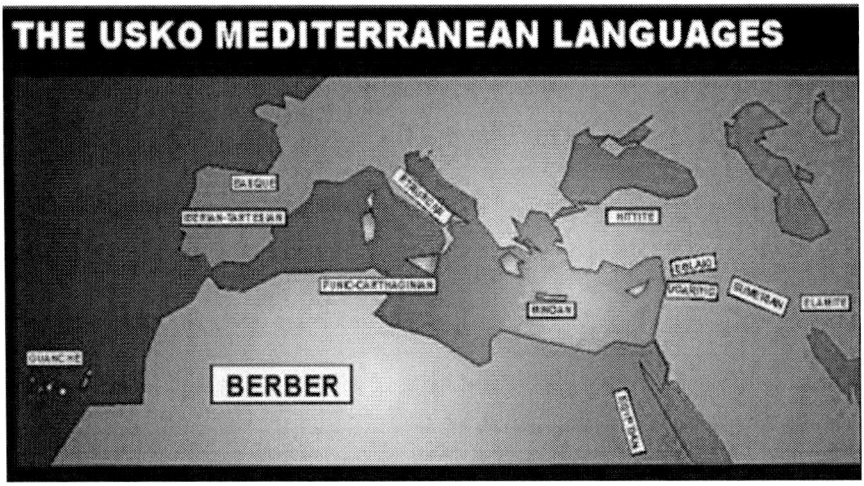

Figure 3. Usko Mediterranean languages *(Arnaiz-Villena et al., 2001)* Amerindian Ethnic Groups HLA Studies.

We analyzed in most Amerindian groups the HLA class II (DRB1 and DQB1) quasi-specific Amerindian allelic lineages (hereafter "alleles" for simplicity) and specific class II HLA haplotypes by using DNA sequencing; in other words, the most frequent HLA alleles and haplotypes in Amerindians which do not exist or exist in very low frequency in other populations, i.e.: genealogy comparisons and Amerindians HLA allele frequencies with those of other First American Natives (Na-Dene, Eskimo and Aleuts) and also those of Asian and Pacific populations with computer programs in order to study the HLA relatedness with peoples most likely to be candidates for First American People ancestors; this would clarify the still unclear peopling of the Americas and the origins of Amerindians, i.e.,: groups of gene frequencies comparisons by using class II HLA allele and haplotype frequencies.

Material and Methods

Population Sample

Eight thousand and fourteen chromosomes from samples of forty seven different populations in several geographical locations (Central and South-East Asia, Australia, Pacific Islands and North, Meso and South America and others) were compared in this study. These samples belong to different ethnic groups, such as Amerindians, Na-Dene, Eskimo, Orientals, Australian aborigines, Polynesians and Melanesians (and controls). Populations are detailed in Table 1. 14,100 chromosomes were used for elaborating dendrogram in Figure 6.

HLA Genotyping

High-resolution HLA class II (DRB1 and DQB1) was performed by PCR-SSOP-Luminex technique (Itoh et al., 2005). This methodology consists of: a) PCR using specific primer pairs of provided by the manufacturers (Luminex Corporation, Austin, TX, USA). All of these primers are 5'-biotined and they are specific to determine sequences of exons 2 and 3 (or only exon 2 for HLA class II) of HLA genes; b) hybridization: products of PCR biotin-labelled were denaturalized at 97°C and then were able to hybridize to complementary DNA probes associated to microbeads; and c) assignation of the HLA alleles: the complex resulting of the hybridization was introduced in Luminex platform, this system identify the fluorescent intensity of fluorophores on each oligobead that has hybridized with the biotin-labelled PCR product. Software of Luminex assigns the HLA alleles for each sample of DNA (Itoh et al., 2005). HLA-DRB1 and -DQB1 allele DNA automated sequencing (ABI PRISM 3700/ ABI PRISM 3730. Applied Biosystems; California) was only done when DNA typing yielded ambiguous results (Arnaiz-Villena et al., 1992).

Aymara (Bolivia), Uros (Peru) and Lamas (Peru) populations were specifically HLA retyped for present work analyses, because of relevance for conclusions.

Statistical Analysis

Statistical analysis was performed with Arlequin v2.0 software kindly provided by Excoffier and Slatkin (Schneider et al., 2000). In summary, this program calculated HLA-DRB1 and -DQB1 allele frequencies, Hardy-Weinberg equilibrium and the linkage disequilibrium between n alleles at n different loci. Their level of significance (p) for 2 x 2 comparisons was determined as previously described (Imanishi et al., 1992b; Imanishi et al., 1992c). In addition, the most frequent complete haplotypes were deduced from: 1) the 2 HLA loci haplotype frequencies (Imanishi et al., 1992b; Imanishi et al., 1992c); 2) the previously described haplotypes in other populations (Imanishi et al., 1992b; Imanishi et al., 1992c); and 3) haplotypes if they appeared in two or more individuals and the alternative haplotype was well defined (Imanishi et al., 1992b; Imanishi et al., 1992c). Phylogenetic trees (dendrograms) were constructed with the HLA-DRB1 allele and HLA class II haplotype frequencies using the Neighbour-Joining (NJ) method (Saitou & Nei, 1987) with the genetic distances between populations (DA) (Nei, 1972), using DISPAN software comprising the programs GNKDST and TREEVIEW (Nei, 1973; Nei et al., 1983). Correspondence analysis in three dimensions and its bidimensional representation was carried out using the VISTA v5.05 computer program (Young & Bann, 1996, *http:/forrest.psych.unc.edu*). Correspondence analysis consists of a geometric technique that may be used for displaying a global view of the relationships among populations according to HLA (or other) allele or haplotype frequencies. This methodology is based on the genetic distances (DA) variance among populations (similar to the classical principal components methodology) and of a statistical visualization of the differences.

RESULTS AND DISCUSSION

The expected and observed gene frequency values for HLA-A, -B, -DRB1 and -DQB1 loci do not differ significantly in studied populations and results are in Hardy-Weinberg equilibrium (results not shown).

We have previously observed that Amerindian HLA frequencies profile (Arnaiz-Villena et al., 2010a, 2010b) does not correlate with either their linguistic branch or their geographical present placements. Clear cut conclusions about Pacific/Amerindian relationships could not be reached, but only some indications (Arnaiz-Villena et al., 2010a, 2010b). However, from Table 3 and Figure 4, it may be drawn that:

1) Aleuts have a different HLA profile (as already found, Moscoso et al., 2008; Rey et al., 2010, see C and D haplotypes)
2) Haplotype F is only found in Amerindians, having Mesoamerican Amerindian its highest frequency. This separates either in space and/or time Amerindians (particularly those from Mesoamerica) from Athabaskans, Asians and Pacific Islanders.
3) Haplotype G relates Pacific Islanders with Amerindians, particularly South American Amerindians, suggesting a more ancient and/or intense gene exchange between Pacific Islanders and Amerindians.

This is also supported because Athabaskans, Yupik (Alaska Eskimos) and Siberians also bear this Pacific Islanders HLA "high" frequency haplotype. Note again that Aleuts lack this haplotype stressing their HLA genetic uniqueness (Moscoso et al., 2008). Gene flow between Amerindian and Pacific Islanders seems to have existed: two Pacific Islanders populations from Samoa and Easter Island share high frequencies (particularly Samoans) with South American Amerindians Mapuches and Uros, together with Taiwan populations. The latter have been postulated by some authors as the first Pacific Prehistoric settlers (Sykes et al., 1995).

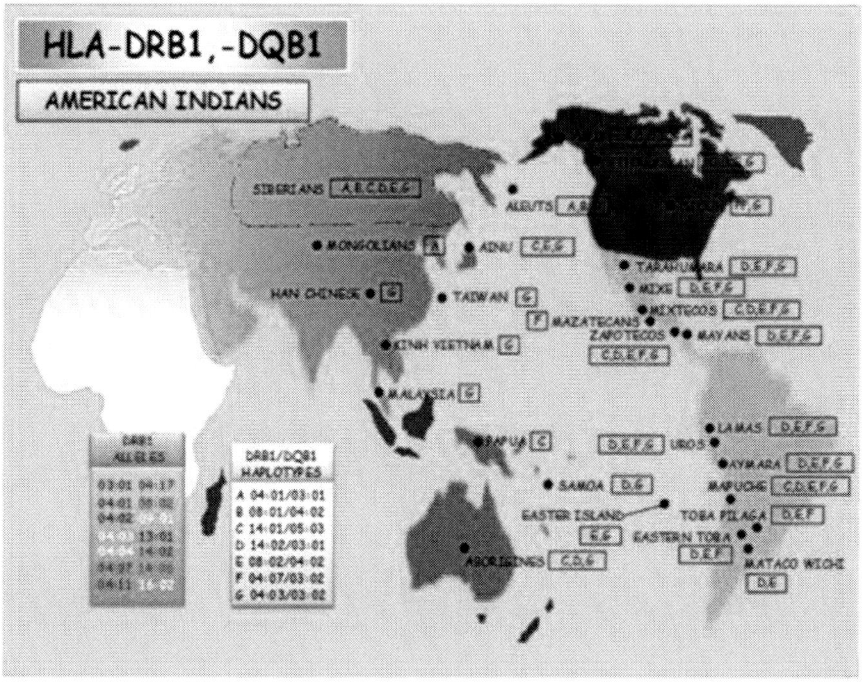

Figure 4. Map showing HLA-DRB1 alleles and HLA-DRB1-DQB1 haplotypes in Siberian, Pacific and Amerindian populations. *Haplotypes are defined by letters from A to G (see Table 3). Alleles are colored according to weather they are present in Amerindians but also in other World populations (black), they are shared by Amerindian and Pacific populations in high frequency (white) or they are mainly present in Amerindians (red). Data from populations were from references stated in Table 1. Amerindian populations living place is colored in green. Reddish to yellow colors mean a gradient of HLA relatedness according to geography. Amerindians are shown as apart in all computer analyses.*

Table 2. Worldwide populations included in this chapter analysis. A Total of 8014 Chromosomes were analyzed

Population	N	Reference	Population	N	Reference
Aborigines	152	(Lester et al., 1995)	Mayans	132	(Gómez-Casado et al., 2003)
Ainu	50	(Bannai et al., 1996)	Mazatecans	89	(Arnaiz-Villena et al., 2000)
Aleuts	85	(Rey et al., 2010)	Mixe	55	(Petzl-Erler et al., 1997)
Athabaskans	62	(Monsalve et al., 1998)	Mixtecans	103	(Petzl-Erler et al., 1997)
Aymaras	102	(Arnaiz-Villena et al., 2005)	Negidal	35	(Uinuk-Ool et al., 2002)
Buryat	25	(Uinuk-Ool et al., 2002)	Nganasan	24	(Uinuk-Ool et al., 2004)
Chukchi	59	(Grahovac et al., 1998)	Nivkhs	32	(Grahovacetal, 1998)
Chuvashians	82	(Arnaiz-Villena et al., 2003)	Papua New Guiñean	65	(Gaoetal, 1992)
Easter Island	48	(Lie et al., 2007)	Quechuas	80	(Martinez-Laso et al., 2006)
Eastern Toba	135	(Cerna et al., 1993)	Samoa	29	(Mack et al., 2000)
Eskimo	35	(Grahovac et al., 1998)	Taiwan	48	(Zimdahl et al., 1999)
Evenks	35	(Grahovac et al., 1998)	Tarahumara	44	(Garcia-Ortiz et al., 2006)
Han Chínese	264	(Trachtenberg et al., 2007)	Tlingit	53	(Imanishi et al., 1992a)
Hottentot	91	(Imanishi et al., 1992a)	Toba Pilaga	19	(Cerna et al., 1993)
Kets	22	(Grahovac et al., 1998)	Tofalar	43	(Uinuk-Ool et al., 2002)
Koryaks	92	(Grahovac et al., 1998)	Todja	22	(Uinuk-Ool et al., 2002)
Khalk Mongolians	202	(Munkhbat et al., 1997)	Tuvinians	197	(Martinez-Laso et al., 2001)
Khoton Mongolians	85	(Munkhbat et al., 1997)	Kinh Vietnam	103	(Vu-Trieu et al., 1997)
Lakota Sioux	302	(Leffell et al., 2004)	Udegeys	23	(Grahovacetal, 1998)
Lamas	83	(Moscoso et al., 2006)	Ulchi	73	(Uinuk-Ool et al., 2002)
Malaysia	74	(Mack et al., 2000)	Uros	105	(Arnaiz-Villena et al., 2009)
Mansi	68	(Uinuk-Ool et al., 2002)	Yupik	252	(Leffell et al., 2002)
Mapuches	104	(Rey et al., 2013)	Zapotecans	75	(Petzl-Erler et al., 1997)
Mataco-Wichi	49	(Cerna et al., 1993)			

N, number of individuals. Population names are in alphabetical order.

Table 3. Shared HLA-DRBl-DQBl Haplotype Frequencies (%) in different Pacific, Siberian and Amerindian Populations

Haplotype HLA-DRBl-DQBl	Samoa	Papua	Easter Island	Taiwan	Ainu	Buryat	Evenks	Ulchi	Negidal	Eskimo	Aleuts	Yupik	Athabaskans
DRB1*04:01-DQB1*03:01 (A)	—	—	—	—	—	8.0	5.7	4.1	9.6	26.2	9.7	22.8	—
DRB1*08:01-DQB1*04:02 (B)	—	—	—	—	—	2.0	—	—	1.4	—	12.5	—	—
DRB1 *'14:01 -DQB1 *05:03 (C)	—	7.6	—	—	20.0	—	—	4.8	1.4	4.4	—	6.7	16.9
DRB1 *14:02-DQB1 *03:01 (D)	1.7	—	—	—	—	—	—	4.1	5.7	20.0	—	22.0	34.7
DRB1 *08:02-DQBl *04:02 (E)	—	—	7.3	—	10.0	—	—	—	4.3	11.3	2.8	13.3	4.8
DRB1*04:07-DQB1*03:02 (F)	—	—	—	—	—	—	—	—	—	—	—	—	—
DRB1 *04:03-DQBl *03:02 (G)	17.2	—	6.3	9.5	3.0	8.0	2.9	—	5.7	—	—	3.0	9.5

Haplotype HLA-DRBl-DQBl	Sioux	Tarahu-mara	Mixe	Mixtecans	Mazate-cans	Zapote-cans	May-ans	Lamas	Uros	Ayma-ra	Quech-uas	Mapu-ches	Toba Pilaga	Eastern Toba	Mata-co Wichi
DRB1*04:01-DQB1 *03:01 (A)	—	—	—	—	—	—	—	—	—	—	—	—	—	—	—
DRB1*08:01-DQB1 *04:02 (B)	—	—	—	—	—	—	—	—	—	—	—	—	—	—	—
DRB1*14:01-DQB1 *05:03 (C)	—	—	—	4.9	—	1.2	—	—	—	—	—	1.0	—	—	—
DRB1*14:02-DQB1 *03:01 (D)	—	27.3	2.9	4.4	—	4.1	1.1	8.4	11.6	10.7	6.5	14.9	7.9	10.4	22.4
DRB1*08:02-DQB1 *04:02(E)	—	35.2	28.0	21.6	—	21.5	15.4	2.4	23.6	22.1	27.9	9.6	10.5	18.9	1.0
DRB1*04:07-DQB1 *03:02(F)	9.3	11.4	17.8	28.9	16.6	9.9	34.8	10.1	2.3	5.4	6.8	2.9	5.3	5.9	—
DRB1*04:03-DQB1 *03:02(G)	2.2	3.4	3.9	3.9	—	5.2	1.4	7.8	11.7	—	2.9	14.9	—	—	—

These Pacific Islands populations (Samoa, Papua and Easter Island) were chosen because of data availability for comparisons and for covering all Pacific Islands distances. Letters in brackets in first column name haplotypes shown in Figure 4.

In conclusion, Mesoamerican Amerindians (related to Maya and Olmec cultures) seem less related with Pacific Islanders than some South Americans Amerindians.

Figure 4 shows in a map how haplotype G relates all Siberians, Eskimo, Athabaskans, Amerindians and Pacific Islanders. Again it is striking its highest frequency in Pacific Islanders and South American Amerindians. If this fact is due to a founder effect then a part of America peopling should have come through the South although other possibilities are open. It is also remarkable that Aleuts do not have this common haplotype (G). This stresses again its unique origin (in space and/or time) from Baikal Lake Area, in Siberia (Moscoso et al., 2008).

Also, this figure shows that DRB1*04:03, 04:04, 09:01 and 16:02 are DRB1 alleles extensively shared with Pacific Islanders; DRB1*03:01, 04:01, 04:02 and 13:01 are present in Amerindians but also in other World populations (Gonzalez-Galarza et al., 2011). HLA DRB 1*04:07, 04:11, 04:17, 08:02, 14:02 and 14:06 are mainly present in Amerindians.

Aymaras and Easter Islanders

There is a little discussion that population of Pacific Islands started in southern China / Taiwan about 5000 years before present (BP). Settlers would first arrived to Melanesia (islands surrounding North and East Australia) and mixed with local Lapita culture people (their origin is not fully explained) (Kayser et al., 2000). Later they would have migrated eastwards to Tonga and Samoa and finally arrived in Easter Island at about 1000 AD (Hunt & Lipo, 2006; Martinson-Wallin & Crockford S.J., 2002). However some prehistoric cultural traits are pointed out to be shared with South American Amerindians: bottle gourd (Green, 2000) and sweet potato (Wallin et al., 2005; Yen, 1974) were cultivated in Easter Island before Europeans officially arrived to Easter Island at the end of 18th Century (Yen, 1974). Fishing (Martinez, 1979), and linguistic and other cultural traits (Jones, 2010; Klar, 2010) strengthen the hypothesis that Easter Island / South Amerindians cultural flow had been established. Titikaka Lake giant stone statues from Tiwanaku culture (like, Monolito Ponce, Monolito Fraile and Monolito Bennet) (Wikipedia, 2013b) are very similar to the giant Easter and other Pacific Islands statues. This led to Thor Heyerdahl to postulate that Amerindians had populated Pacific Islands including Easter Island (Heyerdahl, 1952). Heyerdahl had carried out the Kon-Tiki expedition across Pacific Ocean from Lima (Peru) to Hawaii Archipelago

(3,000 kilometers westwards). He aimed to demonstrate that Amerindians could have colonized Polynesian Islands in ancient times. Particularly, he postulated a link between Titikaka Lake Tiwanaku culture and Easter Island culture. Efforts have been made to find out traces of a possible gene flow between America and Pacific Islands with a remarkable success (Gonçalves et al., 2013).

Tiwanaku culture at Titikaka Lake Area had constructed interesting artefacts like giant statues similar to those found also in Pacific Islands; also, stone holes in temples, which amplify sound, like megaphones (the so-called "ritual ears") are also found both in megalithic-dolmenic Spain (Alberite Dolmen, build up 5000 years ago) and Tiwanaku archaeological (Arnaiz-Villena et al., 2013). Tiwanaku culture was developed by Aymara Amerindians themselves (Arnaiz-Villena et al., 2005) or by both Aymara and other Amerindians previously settled in the area, i.e.: Uros, living on reed-made floating islands at Titikaka Lake, who probably came from Amazon Basin in prehistoric times (Arnaiz-Villena et al., 2009).

Haplotype HLA-A*02-B*39-DRB1*09:01-DQB1*03:03 was found in Aymaras (Arnaiz-Villena et al., 2005) and later shown that its class I A*02:01-B*39:09 part was shared with Easter Islanders (Thorsby, 2012); this haplotype is postulated to be derived from Titikaka Lake living people, Aymara (Arnaiz-Villena et al., 2005). However, Tiwanaku and Easter Island cultures similarities are more significant than genetic coincidences, also more cultural traits are common (Green, 2000; Lawler, 2010; Storey et al., 2007; Storey et al., 2008; Wallin et al., 2005). Ancient gene flow between both Aymara people and Easter Islanders may have been diluted with time and in this particular case may be as useful as cultural shared traits, at present. The same is true for discovering the general relationship between Amerindians and other Pacific Islanders.

Mesoamerican (Mayans, Mixe, Zapotecans, Mixtecans) vs. South American Amerindians

DA genetic distances (Table 4) show a general view: Mesoamericans seem to be less related to Pacific Islanders than South American Amerindians by using two dimension HLA gene frequencies analyses (Table 3).

Table 4. Genetic Distances (DA) Between Pacific Populations and Other Populations (x100) obtained by using HLA-DRB1-DQB1 Haplotype frequencies

Samoa		Papua		Easter Island		Taiwan	
Population	DA (%)	Population	DA (%)	Population	DA (%)	Population	DA (%)
Taiwan	1.55	Ulchi	4.30	Taiwan	3.84	Samoa	1.55
Easter Island	5.88	Evenks	8.10	Samoa	5.88	Evenks	3.80
Buryats	6.72	Taiwan	8.55	Evenks	6.86	Easter Island	3.84
Evenks	6.84	Ainu	8.99	Buryats	8.73	Buryats	5.14
Lamas	8.60	Lakota Sioux	9.57	Lakota Sioux	8.83	Lakota Sioux	5.93
Lakota Sioux	9.13	Easter Island	10.65	Negidal	9.59	Papua	8.55
Negidal	10.62	Mazatecans	12.22	Toba Pilaga	10.05	Lamas	11.06
Mapuches	11.15	Buryats	12.96	Lamas	10.32	Ulchi	11.27
Ulchi	13.36	Samoa	13.43	Papua	10.65	Negidal	11.98
Papua	13.43	Negidal	15.23	Zapotecans	10.90	Mazatecans	13.12
Mataco Wichi	15.01	Mataco Wichi	15.87	Ainu	11.02	Mapuches	16.47
Uros	17.19	Toba Pilaga	16.04	Mapuches	11.95	Mataco Wichi	16.74
Toba Pilaga	17.67	Aleuts	16.75	Quechuas	11.96	Ainu	16.79
Mazatecans	17.76	Lamas	18.83	Uros	12.04	Toba Pilaga	16.90
Ainu	19.10	Eastern Toba	22.62	Ulchi	13.30	Aleuts	17.61
Zapotecans	19.26	Zapotecans	23.71	Eastern Toba	13.43	Zapotecans	20.46
Aleuts	22.01	Aymara	24.43	Aymara	14.23	Uros	21.65
Quechuas	22.28	Mapuches	24.86	Aleuts	14.98	Eastern Toba	23.42
Eastern Toba	23.30	Quechuas	28.13	Mazatecans	15.11	Quechuas	23.62
Aymara	24.94	Uros	31.49	Mataco Wichi	15.95	Aymara	25.21
Athabaskans	26.96	Athabaskans	32.62	Mixe	16.75	Mixe	28.42
Mixe	27.59	Mixe	33.82	Mayans	22.5	Mayans	30.93
Mayans	31.79	Mayans	33.89	Mixtecans	26.48	Athabaskans	34.98
Mixtecans	34.82	Eskimo	34.88	Athabaskans	32.10	Mixtecans	36.60
Yupik	35.60	Mixtecans	35.98	Yupik	33.05	Yupik	40.68
Eskimo	38.58	Yupik	38.32	Eskimo	33.54	Eskimo	41.28
Tarahumara	42.63	Tarahumara	54.20	Tarahumara	35.06	Tarahumara	48.99

However, if genealogy analysis of haplotype G (Table 3, Figure 4, HLA-DRBl*04:03-DQBl-03:02) is used Samoans (17.2%), Taiwanese (9.5%), Mapuches (14.9%), Uros (11.7%) and Lamas (7.8%) cluster together in a high frequency group. Mesoamericans cluster together because of a lower frequency with an apparent separation of Pacific Islanders: Zapotecans (5.2%), Mixe (3.9%), Mixtecans (3.9%) except perhaps with Easter Islanders (6.3%). This HLA haplotype is also found in the Mediterranean area and other parts of the World in low frequency. Isolation and founder effect may have given Samoa people this highest recorded haplotype G frequency.

Mesoamericans tend to cluster together according to HLA allele frequencies in Correspondence and NJ tree analyses Figure 5 (Arnaiz-Villena et al., 2000; Gomez-Casado et al., 2003), and Mayans and Mixe speak clearly Mayan languages and the other Mesoamerican ethnic groups are not very firmly attached to languages different to Mayans (Arnaiz-Villena et al., 2000; Gomez-Casado et al., 2003). Because of this, it is likely that based on these genetic and linguistic evidences, Olmec culture may have given rise to all other Mesoamerican ethnic group cultures. This is supported both on archaeological and genetic data (Arnaiz-Villena et al., 2000; Gomez-Casado et al., 2003). Particularly, Zapotecans constructed Monte Alban, which is one of the most ancient pyramid based complex (Monte Alban, Oaxaca, by 500 years BC) (Arnaiz-Villena et al., 2000) and Mayans imposed their type of languages in a widespread Mesoamerica area (Gomez-Casado et al., 2003). Both Monte Alban and Yucatan first pyramids were constructed by Zapotecans and Mayans respectively by around 500 years BC (Wikipedia, 2013a). Thus, it would seem that Mesoamericans have less relationship with Pacific Islanders.

African-Americans

It is not this chapter topic. However, after 1492 Africans were driven to America as slaves. The first free black Americans were those of San Basilio de Palenque community-Colombia (Arnaiz-Villena et al., 2009b), which nowadays speak a Bantu-Spanish language, unique over the World (Arnaiz-Villena et al., 2009b).

CONCLUSION

America peopling: Amerindians, Na Dene, Eskimo, Aleuts and Pacific Islanders relationship according to HLA autosomal genetic markers

Genetics

1. Frequencies

It is clear that from HLA genetic frequencies and cultural data some conclusions may be drawn about Amerindians and Pacific Islanders relationship. Conclusions are:

1) A three waves' model for America peopling is not supported (Arnaiz-Villena et al., 2000; Arnaiz-Villena et al., 2010a, 2010b).
2) Mesoamericans seem to cluster together according to HLA genes and cultures (Olmecs seem to have given rise to Mayans, Zapotecans, Mixtecans, Mazatecans and Mixe cultures) (Arnaiz-Villena et al., 2000; Arnaiz-Villena et al., 2010a, 2010b; Gomez-Casado et al., 2003).
3) All North, Meso and South American Amerindians cluster together and are separated from other World populations. Amerindians follow little geographical gradient in this HLA frequency analyses, except for Mesoamerican groups (Arnaiz-Villena et al., 2000; Arnaiz-Villena et al., 2010a, 2010b; Gomez-Casado et al., 2003; Rey et al., 2013). 4) Aleuts cluster separately from other groups and may be more related to Baikal Area first Inhabitants and immigrated to North Europe, together with Lapps (Moscoso et al., 2008).

2. Genealogy

1) Amerindians and Pacific Islanders extensively share high frequency and rare (in other parts of World) HLA haplotypes (Table 3: DRB1*04:01-DQB1*03:01, DRB1*08:01-DQB1*04:02, DRB1* 14:01-DQB1*05:03, DRB1*14:02-DQB1*03:01, DRB1*08:02-DQB1*04:02, DRB1*04:07-DQB 1*03:02, DRB1*04:03-DQB 1*03:02; Figure 4: DRB 1*04:03, DRB1*04:04, DRB1*09:01, DRB1*16:02).
2) Cultural and genetic data show an HLA relationship of Easter Islanders and Aymaras (haplotypes A*02:01-B*39:09 and DRB1*08:02-DQB1*0402, Table 2) (Arnaiz-Villena et al., 2005; Thorsby, 2012).
3) Genealogy data in Athabaskans do not discard a two way Amerindian migration through or around Beringia at different times (Arnaiz-Villena et al., 2010a, 2010b).

4) Genealogy data show that Aleuts should be separated in terms of HLA genetics from Eskimo, Athabaskans and Amerindians (Table 2). Also, Aleut language is different from Eskimo and other ancient American languages (Moscoso et al., 2008).

In summary, Pacific peopling seems more complex than thought and genetics by itself may not be a unique tool to uncover it. For instance, it is not easily understandable that Taiwanese and South Chinese populated Pacific Islands for the first time 5000 years BC and in addition found some local aboriginal culture elements like those of Lapita culture (Kayser et al., 2000). This remarks that Pacific Islands were already probably inhabited.

A Multidisciplinary Approach

On the other hand, genetic study of present day populations may not be accurate enough to explain both Pacific (Sykes et al., 1995) and America peopling (Arnaiz-Villena et al., 2010a, 2010b). A solely genetically interpretation of ancient and past peopling may be biased for one or several of the following facts:

1) An ancient founder effect may have disappeared because of a continuous new population admixture effect.
2) Europeans induced a remarkable bottleneck effect in America after 1492: about 85% of American First Inhabitants died because of European borne-diseases during 16th Century (Dobbins, 1993). Amerindians lacked appropriate HLA molecules for starting immune response against new European pathogens. This suggests that Amerindians had a different HLA profile to Europeans before 1492 and that after 16th Century this profile could have farther changed. This putative initial different Amerindian HLA profile may have been more similar to the European one than this that observed at present, because prehistoric isolation may have not been absolute.
3) Stress (like epidemics) induces appearance of new HLA alleles in spermatozoa as demonstrated by single spermatozoa PCR (Huang et al., 1995). A set of new alleles may have appeared after 1492 in Amerindians suffering epidemics of new European borne pathogens.

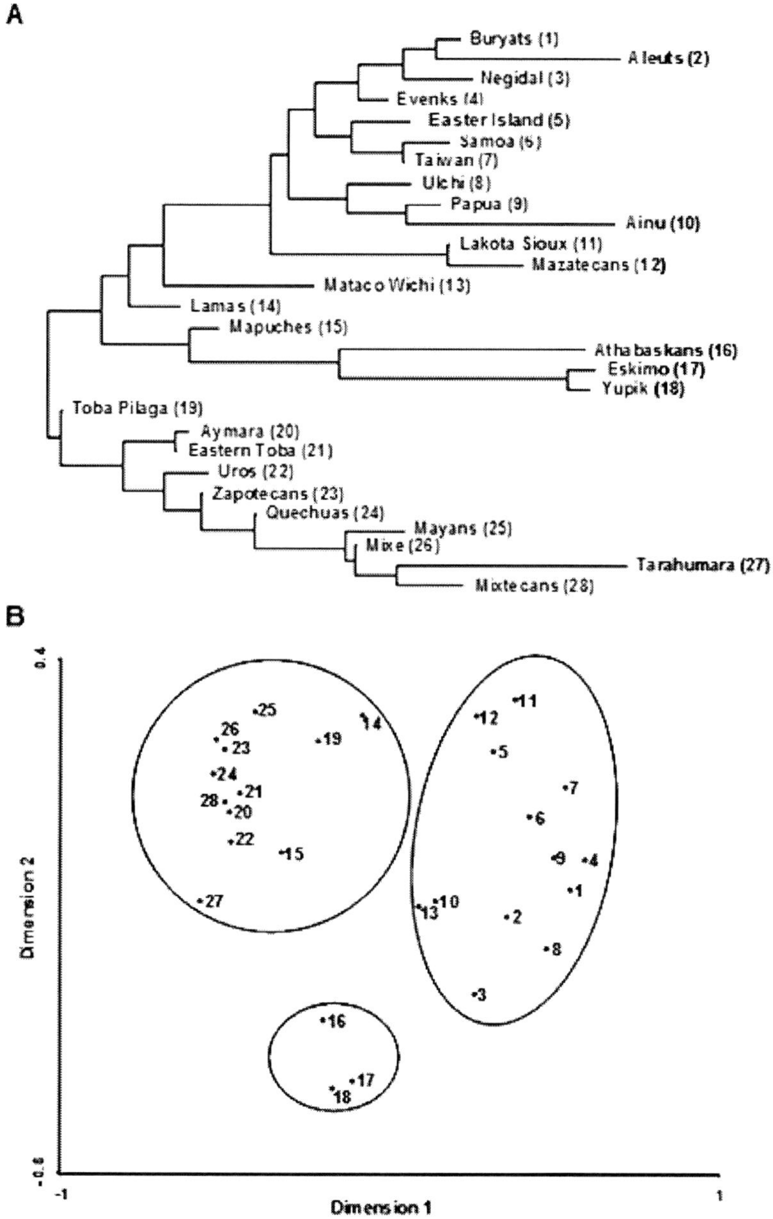

Figure 5. A: Neighbour-Joining (NJ) dendrogram showing relatedness between Amerindians, Pacific and Siberian populations. *Genetic distances between populations (DA) were calculated by using HLA-DRBl-DQBl haplotypes (high resolution). B: Correspondence analysis showing a global view of the relationship between*

Amerindian, Pacific and Siberian populations according to HLA-DRBl-DQBl haplotype frequencies in three dimensions (bidimensional representation). Populations: Buryats (1), Aleuts (2), Negidal (3), Evenks (4), Easter Island (5), Samoa (6), Taiwan (7), Ulchi (8), Papua (9), Ainu (10), Lakota Sioux (11), Mazatecans (12), Mataco Wichi (13), Lamas (14), Mapuches (15), Athabaskans (16), Eskimo (17), Yupik (18), Toba Pilaga (19), Aymara (20), Eastern Toba (21), Uros (22), Zapotecans (23), Quechuas (24), Mayans (25), Mixe (26), Tarahumara (27) and Mixtecans (28). Data from populations were from references stated in Table 1.

Genetic HLA input in America (and reversal output) from Pacific (Arnaiz-Villena et al., 2010a, 2010b) and Atlantic Oceans (Solutreans from Iberia) (Stanford & Bradley, 2012) may explain that Amerindians shared cultural traits with both Pacific (Green, 2000; Lawler, 2010; Storey et al., 2007; Storey et al., 2008; Wallin et al., 2005) and Iberian Solutreans (Stanford & Bradley, 2012) and Iberian megalithic builders (Arnaiz-Villena et al., 2013). Caucasoid Kennewick Man skull found in Columbia River mouth (Morell, 1998), prehistoric Caucasoid skulls found in Brazil (Neves & Pucciareli, 1991) and the fact the most ancient American archaeological sites are in South America (Monte Verde, Pedra Furada) (Dillehay, 1997) points out that American peopling has been more complex than thought (Greenberg et al., 1986). Also it appears to have been varied and complex, and also stratified in space and time. At present, it is difficult to describe it in detail. But all genetic, linguistic, anthropological and cultural findings have to be taking into account.

Finally, there are many questions yet to be answered about America peopling but many more about Pacific Islands peopling. It is clear that a multidisciplinary approach to these problems is necessary, in addition genetics. Results only based on present day genetics may have more than one interpretation of past history and conclusions only drawn from it are all arguable.

Also, results on North American and South American population HLA alleles also support this view. Why nowadays Amerindian populations are altogether different to the rest of the World regarding HLA frequencies is only a matter of speculation (Figure 6): about 80 million early American Natives died during the XVIth century (Dobbins, 1993), mainly due to a lack of appropriate immune response to European-borne diseases, mainly measles, influenza and plague (Dobbins, 1993). This may have shaped the American Native HLA profile by increasing rare HLA alleles able to present new pathogens to T cells (Bodmer and Bodmer., 1978; Slade et al., 1992; Takahata et al., 1990). However, other early North American Natives (non-Amerindians) also suffered many epidemics (Dobbins, 1993) and do not have as different HLA profile from Asians, as Amerindians do.

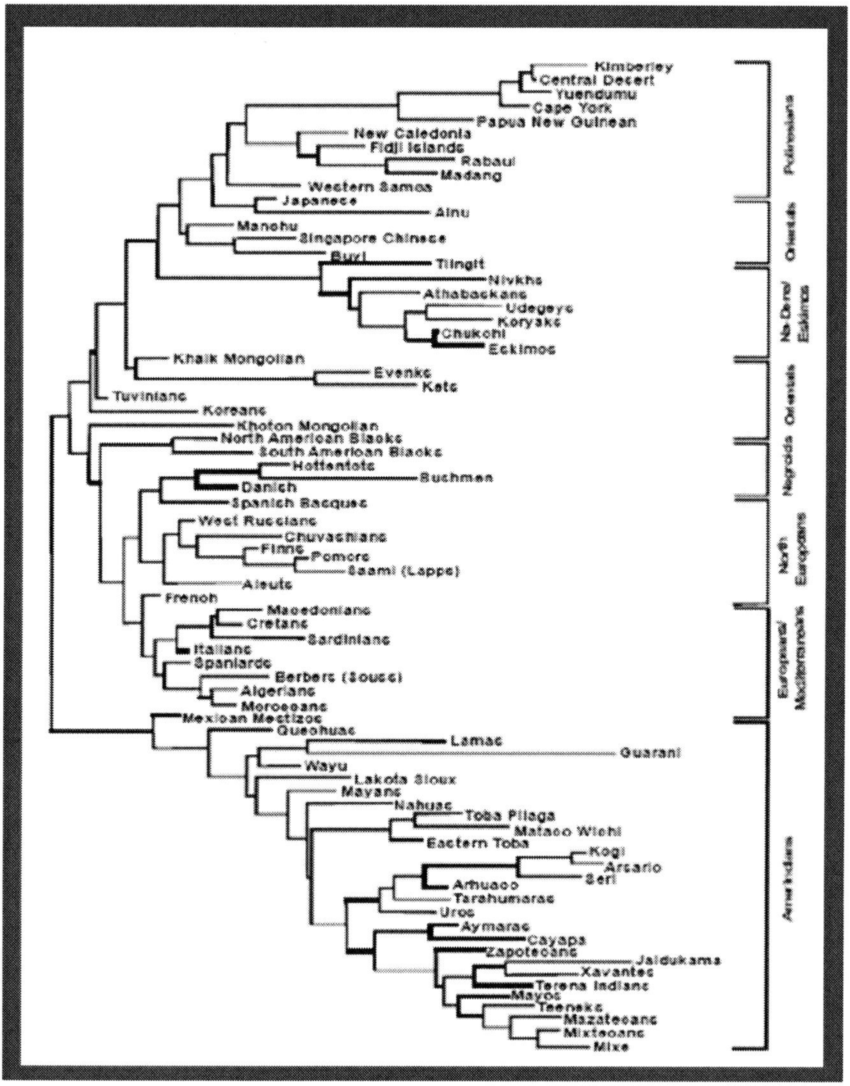

Figure 6. Amerindian relationship within Worldwide populations, including Europeans and Africans. *Neighbor-Joining dendrogram obtained by using HLA-DRB1 allele frequencies. The genetic relatedness among Amerindians, Na-Dene, Eskimos, Asians, Negroids, Europeans and Polynesians are determined by calculating the genetic distances between populations (DA), using HLA-DRB1 allele frequencies. Amerindians cluster together and separated from the rest of the world populations (Arnaiz-Villena et al., 2007; Arnaiz-Villena et al., 2009a; Garcia-Ortiz et al.,2006; Moscoso et al., 2008; Gomez-Casado et al., 2003; Martinez Laso et al., 2006; Moscoso et al., 2006; Vargas-Alarcon et al., 2007).*

Acknowledgments

This work was supported in part by grants from the Spanish Ministry of Health (FISS PI080838 and PI11/00807),FEDER funds and Mutua Madrileña Automovilista grants. The authors report no conflict of interest.

References

Arnaiz-Villena, A., Timón, M., Corell, A., Perez-Aciego, P., Martín-Villa, J.M., et al. (1992). Primary immunodeficiency caused by mutations in the gene encoding the CD3-gamma subunit of the T-lymphocyte receptor. N *Erigi. J. Med.*, 321, 529-533.

Arnaiz-Villena, A., Vargas-Alarcon, G., Granados, J., et al. (2000). HLA genes in Mexican Mazatecans, the peopling of the Americas and the uniqueness of Amerindians. *Tissue Antigens*, 56, 405-416.

Arnaiz-Villena, A., Martinez-Laso, J., Alonso-Garcia, J., (2001). The correlation between languages and genes: The Usko-Mediterranean peoples. *Human Immunology*, 62, 1051-1061.

Arnaiz-Villena, A., Martinez-Laso, J., Moscoso, J. et al. (2003). HLA genes in the Chuvashian population from European Russia: admixture of Central European and Mediterranean populations. *Hum. Biol.*, 75, 375-392.

Arnaiz-Villena, A., Siles, N., Moscoso, J. et al. (2005). Origin of Aymaras from Bolivia and their relationship with other Amerindians according to HLA genes. *Tissue Antigens*, 65, 379-390.

Arnaiz-Villena, A., Moscoso, J., Granados, J., Serrano-Vela, JI., de L Peña, A., et al. (2007). HLA genes in Mayos population from Northeast Mexico. *Mol. Inmunol.*

Arnaiz-Villena, A., Gonzalez-Alcos, V., Serrano-Vela, J.I. et al. (2009a). HLA genes in Uros from Titikaka Lake, Peru: origin and relationship with other Amerindians and worldwide populations. *Int. J. Immunogenet.*, 36, 159-167.

Arnaiz-Villena, A., Reguera, R., Parga-Lozano, C., Abd-El-Fatah-Khalil, S., Monleon, L., et al. (2009b). HLA Genes in Afro-American Colombians (San Basilio de Palenque): The First Free Africans in America. *The Open Immunology Journal*, 2, 59-66

Arnaiz-Villena, A., Parga-Lozano, C, Moreno, E., Areces, C, Rey, D., et al. (2010a). The Origin of Amerindians and the Peopling of the Americas

According to HLA Genes: Admixture with Asian and Pacific People. *Curr. Genomics*, 11, 103-114.

Arnaiz-Villena, A., Areces, C, Gómez-Prieto, P., Parga-Lozano, P., Moreno, E., et al. (2010b). The peopling of the Americas: a complex issue for Amerindian, Na-Dene, Aleut and Eskimo first inhabitants. International *Journal of Modern Antrophology*, 3, 65-79.

Arnaiz-Villena, A., Alonso-Rubio, J., Ruiz-del-Valle, V., (2013). Tiwanaku (Titikaka Lake, Bolivia) and Alberite Dolmen (Southern Spain) ritual "ears." Celtic, Iberian, Aymara and Basque languages. *Int. J. Mod. Anthrop.*, 6, 61-76.

Arnaiz-Villena, A, Areces, C, Enriquez-de-Salamanca, M., et al. (2014). Pacific islanders and Amerindian relatedness according to HLA autosomal genes. *International Journal of Modern Anthropology*, 7, 44-67.

Bannai, M., Tokunaga, K., Imanishi, T., et al. (1996). HLA class II alleles in Ainu living in Hidaka District, Hokkaido, northern Japan. *Am. J. Phys. Anthropol.*, 101, 1-9.

Bakker, P., Martin, H., Hualde, M., (1991) Basque pidgins in Iceland and Canada. Ed.: Universidad País Vasco, Seminario Filología Julio Urquijo, Bilbao, Spain.

Bodmer, W.F., Bodmer, J.G., (1978). Evolution and function of the HLA system. *Br. Med. Bull.*, 34, 309-316.

Bruges-Armas, J., Martinez-Laso, J., Martins, B., et al. (1999). HLA in the Azores Archipelago: possible presence of Mongoloid genes. *Tissue Antigens*, 54, 349-359.

Cavalli-Sforza, L.L., Menozzi, P., Piazza, A., (1994). The history and geography of human genes. Princeton: Princeton University Press.

Cerna, M., Falco, M., Friedman, H., et al. (1993). Differences in HLA class II alleles of isolated South American Indian populations from Brazil and Argentina. *Hum. Immunol.*, 37, 213-220.

Dillehay, T.D., (1997). Monte Verde: A late Pleistocene settlement in Chile. Washington DC: Smithsonian Institution Press.

Dobbins, F., (1993). Disease transfer contact. *Annu. Rev. Anthropol.*, 22, 273-291.

Gao, X., Bhatia, K., Trent, R.J., Serjeantson, S.W., (1992). HLA-DR, DQ nucleotide sequence polymorphisms in five Melanesian populations. *Tissue Antigens*, 40, 31-37.

Garcia-Ortiz, J.E., Sandoval-Ramirez, L., Rangel-Villalobos, H., et al. (2006). High-resolution molecular characterization of the HLA class I and class II in the Tarahumara Amerindian population. *Tissue Antigens*, 68, 135-146.

Goebel, T., Waters, M.R., O'Rourke, D.H., (2008). The late Pleistocene dispersal of modern humans in the Americas. *Science*, 319, 1497-1502.

Gómez-Casado, E., Martinez-Laso, J., Moscoso, J. et al. (2003). Origin of Mayans according to HLA genes and the uniqueness of Amerindians. *Tissue Antigens*, 61, 425-436.

Gonçalves, V.F., Stenderup, J., Rodrigues-Carvalho, C. et al. (2013). Identification of Polynesian mtDNA haplogroups in remains of Botocudo Amerindians from Brazil. *Proc. Nati. Acad. Sci. USA*, 110, 6465-6469.

Gonzalez-Galarza, F.F., Christmas, S., Middleton, D., Jones, A.R., (2011). Allele frequency net: a database and online repository for immune gene frequencies in worldwide populations. *Nucleic Acids Res.,* 39, (Database issue) D913-D919.

Grahovac, B., Sukernik, R.I., O'huigin, C. et al. (1998). Polymorphism of the HLA class II loci in Siberian populations. *Hum. Genet.*, 102, 27-43.

Green, R.C., (2000). Shorter communication: A range of disciplines support a dual origin for the bottle gourd in the Pacific. *The Journal of the Polynesian Society*, 106, 191-198.

Greenberg, J.H., Turner, C.G., Zegura, S.L., (1986). The settlement of the Americas: a comparison of the linguistic, dental and genetic evidence. *Curr. Anthropol.*, 27, 477-498.

Heyerdahl, T., (1952). American Indians in the Pacific. The theory behind the Kon-Tiki expedition. London, UK, Alien and Unwin.

Holden, C, (1999). Were Spaniards among the first Americans? *Science*, 286, 1467-1468.

Horai, S., Kondo, R., Nakagawa-Hattori, Y., Hayashi, S., Sonoda, S., (1993). Peopling of the Americas, founded by four major lineages of mitochondrial DNA. *Mol. Biol. Evol.*, 10, 23-47.

Huang, M.M., Erlich, H.A., Goodman, M.F., Arnheim, N., (1995). Analysis of mutational changes at the HLA locus in single human sperm. *Hum. Mutat.*, 6, 303-310.

Hunt, T.L., Lipo, C.P., (2006). Late colonization of Easter Island. *Science,* 311, 1603-1606.

Imanishi, T., Akaza, T., Kimura, A., Tokunaga, K., Gojobori, T., (1992a). Allele and haplotype frequencies for HLA and complement loci in various ethnic groups. In: Tsuji, K., Aizawa, M., Sasazuki, T. ed. HLA 1991. Oxford: *Oxford University Press*, 1065-1220.

Imanishi, T., Akaza, T., Kimura, A., Tokunaga, K., Gojobori, T., (1992b). Estimation of allele and haplotype frequencies for HLA and complement loci. In: Tsuji, K., Aizawa, M., Sasazuki, T. ed. HLA 1991. Oxford: *Oxford University Press*, 76-79.

Imanishi, T., Akaza, T., Kimura, A., Tokunaga, K., Gojobori, T., (1992c). Genetic relationships among various human populations indicated by MHC polymorphisms. In: Tsuji, K., Aizawa, M., Sasazuki, T. ed. HLA 1991. Oxford: *Oxford University Press*, 627-632.

Itoh, Y., Mizuki, N., Shimada, T., et al. (2005). High-throughput DNA typing of HLA-A, -B, -C, and -DRB1 loci by a PCR-SSOP-Luminex method in the Japanese population. *Immunogenetics*, 57, 717-729.

Jones, T.L., (2010). Polynesians in the new world: the Chumash connection and beyond. In: Wallin, P., Martinson-Wallin, H. ed. Migration, identity and cultural heredity. Selected papers from the 7th Int. Conf. on Easter Island and the Pacific. Visby, Sweden: *Gotland University Press*, 307-318.

Karafet, T., Zegura, S.L., Vuturo-Brady, J., et al. (1997). Y chromosome markers and Trans-Bering Strait dispersals. *Am. J. Phys. Anthropol.*, 102, 301-314.

Karafet, T.M., Zegura, S.L., Posukh, O., et al. (1999). Ancestral Asian source(s) of new world Y-chromosome founder haplotypes. *Am. J. Hum. Genet.*, 64, 817-831.

Kayser, M., Brauer, S., Weiss, G., et al. (2000). Melanesian origin of Polynesian Y chromosomes. *Curr. Biol.*, 10, 1237-1246.

Klar, K.A., (2010). Linguistic evidence for prehistoric Polynesian-American contact. In: Wallin, P., Martinsson-Wallin, H. ed. Migration, identity and cultural heredity. Selected papers from the 7th Int. Conf. on Easter Island and the Pacific. Visby, Sweden: *Gotland University Press*, 319-324.

Kolman, C.J., Sambuughin, N., Bermingham, E., (1996). Mitochondrial DNA analysis of Mongolian populations and implications for the origin of New World founders. *Genetics*, 142, 1321-1334.

Lawler, A., (2010). Beyond Kon-Tiki: did Polynesians sail to South America? *Science*, 328, 1344-1347.

Leffell, M.S., Fallin, M.D., Erlich, H.A., et al. (2002). HLA antigens, alleles and haplotypes among the Yupik Alaska natives: report of the ASHI Minority Workshops, Part II. *Hum. Immunol.*, 63, 614-625.

Leffell, M.S., Fallin, M.D., Hildebrand, W.H., Cavett, J.W., Iglehart, B.A, (2004). HLA alleles and haplotypes among the Lakota Sioux: report of the ASHI minority workshops, part III. *Hum. Immunol.*, 65, 78-89.

Leon-S, F.E., Ariza-Deleon, A., Leon-S, M.E., Ariza, C., (1996). Peopling the Americas. *Science*, 273, 723-725.

Lester, S., Cassidy, S., Humphreys, L, et al. (1995). Evolution in HLA-DRBl and major histocompatibility complex class II haplotypes of Australian aborigines. Definition of a new DRB1 allele and distribution of DRB1 gene frequencies. *Hum. Immunol.*, 42, 154-160.

Lie, B.A., Dupuy, B.M., Spurkland, A., Fernandez-Viña, M.A., Hagelberg, E., (2007). Molecular genetic studies of natives on Easter Island: evidence of an early European and Amerindian contribution to the Polynesian gene pool. *Tissue Antigens*, 69, 10-18.

Mack, S.J., Bugawan, T.L., Moonsamy, P.V., et al. (2000). Evolution of Pacific/Asian populations inferred from HLA class II allele frequency distributions. *Tissue Antigens*, 55, 383-400.

Martínez, A.L., (1979). 9700 years of maritime subsistence of the Pacific: an analysis by means of bioindicators in the North of Chile. *American Antiquity*, 44, 309-324.

Martinez-Laso, J., Sartakova, M., Allende, L., et al. (2001). HLA molecular markers in Tuvinians: a population with both Oriental and Caucasoid characteristics. *Ann. Hum. Genet.*, 65, 245-261.

Martinez-Laso, J., Siles, N., Moscoso, J., et al. (2006). Origin of Bolivian Quechua Amerindians: their relationship with other American Indians and Asians according to HLA genes. *Eur. J. Med. Genet.*, 49, 169-185.

Martinson-Wallin, H., Crockford, S.J., (2002). Early settlement of Rapa Nui (Easter Island). *Asían Perspectives*, 40, 244-278.

Matisoo-Smith, E.A., Ramirez, J.M., (2010). Human skeletal evidence of Polynesian presence in South America? Metric analyses of six crania from Mocha Island, Chile. *Journal of Pacific Archaeology*, 1, 76-88.

Merriwether, D.A., Hall, W.W., Vahlne, A., Ferrell, R.E., (1996). mtDNA variation indicates Mongolia may have been the source for the founding population for the New World. *Am. J. Hum. Genet.*, 59, 204-212.

Monsalve, M.V. Edin, G., Devine, D.V., (1998). Analysis of HLA class I and class II in Na-Dene and Amerindian populations from British Columbia, Canada. *Hum. Immunol.*, 59, 48-55.

Morell, V., (1998). Kennewick Man's trials continue. Science, 280, 190-192.

Moreno-Mayar, J.V., Rasmussen, S., Seguin-Orlando, A., et al. (2014). Genome-wide Ancestry patterns in Rapanui suggest pre-European admixture with Native Americans. *Cur. Biol.*, 24, 2518-2525.

Moscoso, J., Crawford, M.H., Vicario, J.L., et al. (2008). HLA genes of Aleutian Islanders living between Alaska (USA) and Kamchatka (Russia) suggest a possible southern Siberia origin. *Mol. Immunol.*, 45, 1018-1026.

Moscoso, J., Seclen, S., Serrano-Vela, J.I., et al. (2006). HLA genes in Lamas Peruvian-Amazonian Amerindians. *Mol. Immunol.*, 43, 1881-1889.

Mulligan, C.J., Kitchen, A., Miyamoto, M.M., (2008). Updated three-stage model for the peopling of the Americas. *PLoS One*, 3, e3199.

Munkhbat, B., Sato, T., Hagihara, M., et al. (1997). Molecular analysis of HLA polymorphism in Khoton-Mongolians. *Tissue Antigens*, 50, 124-134.

Nei, M., (1972). Genetic distances between populations. *Am. Nat.*, 106, 283.

Nei, M., (1973). Analysis of gene diversity in subdivided populations. *Proc. Nati. Acad. Sci. USA*, 70, 3321-3323.

Nei, M., Tajima, F., Tateno, Y., (1983). Accuracy of estimated phylogenetic trees from molecular data II. Gene frequency data. *J. Mol. Evol.*, 19, 153-170.

Neves, W.A., Pucciareli, H.M., (19919. Morphological affinities of the First Americans: an exploratory analysis based on early South American remains. *J. Hum. Evol.*, 21, 261-273.

Novick, G.E., Novick, C.C., Yunis, J. et al. (1998). Polymorphic Alu insertions and the Asian origin of Native American populations. *Hum. Biol.*, 1, 23-39.

Parham, P., Ohta, T., (1996). Population biology of antigen presentation by MHC class I molecules. *Science,* 272, 61-1 A.

Patricia, M. and Bikal, P.M., (1990). Timelines: a Phoenician Fable. *Archaeology*. Jan-Feb 1990.

Petzl-Erler, M.L., Gorodezky, C., Layrisse, Z., et al. (1997). Anthropology report for the Latin-American Region: Amerindian and admixture populations. In: Charron, D. ed. Paris: EDK, 337-345.

Reidla, M., Kivisild, T., Metspalu, E., et al. (2003). Origins and diffusion of mtDNA haplogroup X. *American Journal of Human Genetics*, 73, 1178-90.

Rey, D., Areces, C, Parga-Lozano, C, Gómez-Prieto, P., Crawford, M.H., (2010). HLA genes in populations of the Aleutian islands. *Hum. Biol.*, 82, 737-744.

Rey, D., Parga-Lozano, C, Moscoso, J., et al. (2013). HLA genetic profile of Mapuche (Araucanian) Amerindians from Chile. *Mol. Biol. Rep.,* 40, 4257-4267.

Rodriguez, A.F., (2015). Megalithic constructions discovered in Azores, Portugal. *Archaeological Discovery*, 3, 51-61.

Ruhlen, M., (1991). A guide of the world's languages: Volume I, Classification. Stanford University Press, Stanford (USA).

Saitou, N., Nei, M., (1987). The neighbor-joining method: a new method for reconstructing phylogenetic trees. *Mol. Biol. Evol.*, 4, 406-425.

Santos, F.R., Pandya, A., Tyler-Smith, C, et al. (1999). The central Siberian origin for native American Y chromosomes. *Am. J. Hum. Genet.*, 64, 619-628.

Santos, F.R., Rodriguez-Delfin, L., Pena, S.D., Moore, J., Weiss, K.M., (1996). North and South Amerindians may have the same major founder Y chromosome haplotype. *Am. J. Hum. Genet.*, 58, 1369-1370.

Schneider, S., Roessli, D., Excoffier, L., (2000). ARLEQUÍN: a software for population genetics database analysis 2.0.

Slade, R.W., McCallum, H.I.,(1992). Overdominance vs frequency-dependent selection at MHC loci. *Genetics*. 132: 861-862.

Stanford, D.J., Bradley, B.A., (2012). Across Atlantic ice: the origins of America's Clovis Culture. Berkeley, USA: University of California Press.

Storey, A.A., Quiroz, D., Ramirez, J.M., et al. (2008). Pre-Columbian chickens, dates, isotopes, and mtDNA. *Proc. Nat. Acad. Sci. USA* 105: E99.

Storey, A. A., Ramirez, J.M., Quiroz, D., et al. (2007). Radiocarbon and DNA evidence for a pre-Columbian introduction of Polynesian chickens to Chile. *Proc. Nati. Acad. Sci. USA*, 104, 10335-10339.

Sykes, B., Leiboff, A., Low-Beer, J., Tetzner, S., Richards, M., (1995). The origins of the Polynesians: an interpretation from mitochondrial lineage analysis. *Am. J. Hum. Genet.*, 57, 1463-1475.

Takahata, N., Nei, M., (1990). Allelic genealogy under overdominant and frequency-dependent selection and polymorphism of major histocompatibility complex loci. *Genetics* 124: 967-978.

Thorsby, E., (2012). The Polynesian gene pool: an early contribution by Amerindians to Easter Island. Philos. *Trans R. Soc. Lond B Biol. Sci.*, 367, 812-819.

Torroni, A., Sukernik, R.I., Schurr, T.G., et al. (1993). mtDNA variation of aboriginal Siberians reveals distinct genetic affinities with Native Americans. *Am. J. Hum. Genet.,* 53, 591-608.

Trachtenberg, E., Vinson, M., Hayes, E., et al. (2007). HLA class I (A, B, C) and class II (DRB1, DQA1, DQBl, DPB1) alleles and haplotypes in the Han from southern China. *Tissue Antigens,* 70, 455-463.

Uinuk-Ool, T.S., Takezaki, N., Derbeneva, O.A, Volodko, N.V., Sukernik, R.I., (2004). Variation of HLA class II genes in the Nganasan and Ket, two aboriginal Siberian populations. *Eur. J. Immunogenet.*, 31, 43-51.

Uinuk-Ool, T.S., Takezaki, N., Sukernik, R.I., Nagl, S., Klein, J., (2002). Origin and affinities of indigenous Siberian populations as revealed by HLA class II gene frequencies. *Hum. Genet.*, 110, 209-226.

Untermann, J., (2006). Sobre la existencia de lenguas de sustrato en la Península (Ibérica?). Revista Real Academia de Cultura Valenciana (ELEA), Valencia (Spain).

Vargas-Alarcon, G., Moscoso, J., Martinez-Laso, J., Rodriguez-Perez, JM., Flores Domínguez, C., et al. (2007). Origin of Mexican Nahuas (Aztecs) according to HLA genes and their relationships with worldwide populations. *Mol. Immunol.*, 44,747-755.

Von Petzinger, G. (2009). Making the abstract concrete: the place of geometric signs in French upper Palaeolithic parietal art. Theses (Anthropology), ETD (Electronic Theses and Dissertations)

Vu-Trieu, A., Djoulah, S., Tran-Thi, C, et al. (1997). HLA-DR and -DQB1 DNA polymorphisms in a Vietnamese Kinh population from Hanoi. *Eur. J. Immunogenet.*, 24, 345-356.

Wallace, D.C., Torroni, A., (1992). American Indian prehistory as written in the mitochondrial DNA: a review. *Hum. Biol.*, 64, 403-416.

Wallin, P., Stevenson, C, Ladefoged, T., (2005). Sweet potato production on RapaNui. In: Ballard, C, Brown, P., Bourke, M., Harwood, T. ed. The sweet potato in Oceania: a reappraisal. *Ethnology Monographs* 19, Oceania Monograph 56. Pittsburg, USA: University of Pittsburg, 85-88.

Wikipedia. (2013a). Monte Albán. Wikipedia, the free encyclopedia. *http://en.wikipedia. org/wiki/Monte_Albán*. Accessed 26 November 2013a.

Wikipedia. (2013b). Tiahuanaco. Wikipedia, la enciclopedia libre. *http://es.wikipedia.org/ wiki/Tiahuanaco*. Accessed 26 November 2013b.

Yen, D.E., (1974). The sweet potato and Oceania. Bulletin no. 236. Honolulú, USA: Bernice P. Bishop Museum,

Young, F.W., Bann, C.M., (1996). A visual statistics system. In: Stine, R. A., Fox, J. ed. Statistical Computing Environments for Social Researches. London: Sage Publications, 207-236.

Zink, D., (1980) The ancient stones speak. *Ed. Plume Books*. New York, USA.

Zimdahl, H., Schiefenhovel, W., Kayser, M., Roewer, L., Nagy M., (1999). Towards understanding the origin and dispersal of Austronesians in the Solomon Sea: HLA class II polymorphism in eight distinct populations of Asia-Oceania. *Eur. J. Immunogenet.*, 26, 405-416.

In: Native Americans
Editor: Jerome Mendoza
ISBN: 978-1-63483-184-0

Chapter 3

THE RELATIONSHIP BETWEEN ETHNIC IDENTITY, RELIGIOUS IDENTITY, AND SELF-ESTEEM IN A SAMPLE OF LUMBEE YOUTH*

Katie L. Cotter[1], PhD, MSW, Caroline B. R. Evans[1], PhD, MSW and Paul R. Smokowski[1,2], PhD, MSW
[1]University of North Carolina – Chapel Hill
North Carolina Academic Center for Excellence, NC, US
in Youth Violence Prevention
[2]Arizona State University, AZ, US

ABSTRACT

Although a strong ethnic identity has been associated with self-esteem, few studies have used longitudinal data to examine mediational

* This study was funded through a cooperative agreement from the United States Centers for Disease Control and Prevention's National Center for Injury Prevention and Control (5 U01 CE001948) to the North Carolina Academic Center for Excellence in Youth Violence Prevention (NC-ACE). Please direct correspondence to: Paul Smokowski, Ph.D., University of North Carolina at Chapel Hill, School of Social work, CB # 3550, 325 Pittsboro Street, Chapel Hill, NC 27599-3550; smokowsk@email.unc.edu. Fax: 919-962-1698; Phone: 919-428-5716.

mechanisms that account for this relationship, especially among American Indian youth. This study investigated religious identity as a mediator of the relationship between ethnic identity and self-esteem. Structural Equation Modeling was used to investigate mediation pathways from ethnic identity to self-esteem in a sample (N = 1,571) of Lumbee middle school students. Results showed that the relationship between Year 1 ethnic identity and Year 3 self-esteem was partially mediated through Year 2 religious identity. Implications were discussed.

Keywords: Ethnic identity, religious identity, self-esteem

Adolescence is a tumultuous period marked by exploration and a desire for autonomy. A central goal of this stage is identity formation (Erikson, 1950) and adolescents strive to solidify a sense of self by spending increasing time with peers and exploring new activities. In addition to forming new relationships and experimenting with activities, adolescents also frequently explore their ethnic identity (Nishina, Bellmore, Witkow, & Nylund-Gibson, 2010). The concept of 'ethnic identity' refers to one's self-identification with an ethnic group and their emotional responses to that group (Bernal & Knight, 1993). Specifically, ethnic identity is the degree to which an individual embraces his or her ethnicity (Phinney, Horenczyk, Liebkind, & Vedder, 2001). A strong ethnic identity serves as a protective factor as it provides youth with a sense of group membership and belonging. Indeed, a substantial body of research confirms the positive association between ethnic identity and self-esteem (e.g., Blash & Unger, 1995; Corenblum& Armstrong, 2012; Phinney & Chavira, 1992; Phinney et al., 2001; Umaña-Taylor & Updegraff, 2007).

Ethnic identity may be a particularly salient protective factor among American Indian youth. Despite high rates of poverty, social barriers, and a long history of forced assimilation and discrimination (Stannard, 1992), many American Indians have maintained strong cultural identities. That is, cultural practices, beliefs, and tribal affiliations remain a significant part of the lives of American Indians. The role of ethnic identity is particularly pertinent to the Lumbee tribe of the south-central region of North Carolina. The Lumbee tribe has historically struggled to obtain legal recognition as an American Indian tribe. Although the State of North Carolina has formally recognized the tribe, the Lumbee people remain unrecognized by the U.S. federal government (Bryant & LaFromboise 2005; Lowery 2010). The Lumbee Act of 1956 granted the Lumbee recognition as "Indian," but denied them the full benefits of tribal recognition. The Lumbee's longstanding struggle for recognition and equality contributes to the importance of ethnic identity among this group (Bryant & LaFromboise 2005; Dial & Elliades, 1975), as ethnic identity serves to strengthen ties to important cultural practices in the face of ongoing persecution.

Religion and connection with the natural world are also important elements of Lumbee culture. The majority of the Lumbee's religious practices align with the Southern Baptist faith (Bryant & LaFromboise 2005) and church is a part of the Lumbee's daily lives. Although the relationship between ethnic identity and self-esteem is well established, researchers often fail to take into account other factors, such as religious identity, that might impact this relationship. Religious identity refers to the meanings individuals ascribe to religion and to individual levels of religiosity (e.g., importance of religion, religious beliefs, attendance of religious gatherings; Keyes & Reitzes, 2007). Historically, religion has been investigated as an important component of ethnic identity (Phinney, 1990) as the two constructs are closely tied (Bankston & Zhou, 1995; Chong, 1998; Garroutte et al., 2009; Lopez, Huynh, & Fuligni, 2011). Yet few researchers have investigated the relationship between ethnic identity, religious identity, and self-esteem, especially among Lumbee youth. Recently, however, in a sample that included Lumbee, African American, Latino/Hispanic, and White youth, researchers found that self-esteem was significantly and positively associated with ethnic identity until religious identity was added into the model (Evans, Smokowski, & Cotter, 2014). Religious identity accounted for much of the relationship between ethnic identity and self-esteem. The aim of the current study is to further examine the relationship among these three constructs specifically for Lumbee youth.

Social Identity Theory

The term 'social identity' refers to the part of one's sense of self that is derived from membership in a social group (Tajfel, 1982). According to social identity theory, group membership plays a central role in identity formation (Tajfel & Turner, 1979). Given that ethnic identity is part of social identity (Phinney & Ong, 2007), this theory is frequently used as a theoretical framework in studies of ethnic identity (e.g., Corenblum & Armstrong, 2012; Phinney, 1990; Phinney & Ong, 2007). Group membership fosters a positive self-concept by providing individuals with a sense of belonging (Tajfel &

Turner, 1979). Further, when members define their group positively in relation to others, group membership serves to maintain self-esteem (Bourhis & Hill, 1982).A strong ethnic identity indicates the presences of positive feelings about one's ethnic group and indicates a feeling of belonging, which in turn might serve to foster self-esteem.

Religion is another component of social identity (Tajfel, 1982) as religious gatherings foster a sense of community and group membership. Religious identity refers, in part, to the importance an individual places on religion. As many spiritual groups are dominated by a specific ethnic group (Dougherty, 2003), religious identity and ethnic identity are closely tied (Garroutte et al., 2009; Lopez et al., 2011). In fact, it is possible that ethnic identity influences the religious group with which individuals affiliate. For example, as individuals interact and form bonds with members of their own ethnic group, they might discuss religion, thereby bolstering religious identity. Thus, ethnic identity might influence religion and both constructs consequently lead to a sense of group membership and increased self-esteem. These relationships remain unexplored among Lumbee youth.

THE CONNECTION BETWEEN ETHNIC IDENTITY AND SELF-ESTEEM

Although the connection between ethnic identity and self-esteem is well researched, results underscore conflicting views of the direction and strength of this relationship (see Phinney, 1990), especially across race/ethnicity. Further, there is minimal research examining this relationship in Lumbee adolescents and the existing literature focused on American Indian youth is inconsistent. Generally, there is a positive association between ethnic identity and self-esteem (Blash& Unger, 1995; Corenblum& Armstrong, 2012; Phinney &Chavira, 1992; Phinney et al., 2001; Umaña-Taylor & Updegraff, 2007). It is possible that high self-esteem fuels the investigation of ethnic identity by giving youth the confidence to challenge ethnic stereotypes or seek out positive ethnic role models. Conversely, it could be that ethnic identity increases positive feelings about oneself by instilling in youth knowledge and pride about hardships overcome by one's ethnic group (Phinney &Chavira, 1992).

Many existing studies have examined how ethnic identity predicts self-esteem and have found that ethnic identity is a statistically significant predictor of self-esteem in diverse samples (e.g., Blash & Unger, 1995; Evans et al., 2014; Martinez & Dukes, 1997; Phinney & Chavira, 1992; Phinney, Cantu, & Kurtz, 1997). However, investigations among American Indian youth have led to mixed findings: in one sample of Navajo youth, Jones and Galiher (2007) identified a positive relationship between ethnic identity and self-esteem, whereas other studies of American Indian youth have reported an inverse relationship between these constructs (i.e., in a sample of Native Aboriginal Canadian youth, Corenblum & Armstrong, 2012; in a sample of Lakota/Dakota Sioux American Indian youth, Pittenger, 1998). Based on these inconsistent findings, further research is warranted among Lumbee Indian youth. Further, it is possible that other factors, such as religious identity, play a role in the relationship between ethnic identity and self-esteem. The concept of religious identity has remained unmeasured in many investigations.

The Connection between Ethnic Identity and Religious Identity

Religion is considered a key variable in the study of ethnic identity (Phinney, 1990) and research highlights the role of religious beliefs in ethnic identity maintenance (Chong, 1998). For example, in a sample of 477 Latino/Hispanic, Caucasian, and Asian youth, ethnic and religious identities were stable over high school and changes in one identity often prompted changes in the other (Lopez et al., 2011). The connection between ethnic identity and religion might be especially strong for Lumbee youth given the prominent role that religion plays in their lives and the fact that church congregations are often dominated by certain ethnic groups (Dougherty, 2003). Social supports, such as religious gatherings, might assist American Indians in dealing with historical trauma by providing a sense of solidarity and fostering community and resilience. Indeed, in a sample that included Lumbee, African American, Latino/Hispanic, and White youth, researchers found that self-esteem was significantly and positively associated with ethnic identity until religious identity was added into the model (Evans et al., 2014).

THE CONNECTION BETWEEN RELIGIOUS IDENTITY AND SELF-ESTEEM

Overall, research on the relationship between religious identity and self-esteem is mixed. One study of 13,317 African American, Asian American, Caucasian, Latino/Hispanic, and American Indian adolescents found that participation in religious activities (e.g., attending services, participating in youth activities, praying) and a belief in the importance of religion were associated with increased self-esteem for all groups except American Indians (Le, Tov, & Taylor, 2007). In a diverse sample of 904 adolescents from Canada, Britain, Hong Kong, and the Philippines, there was a positive and statistically significant association between self-esteem and religious identity (Bagley & Mallick, 1997). Other researchers confirm the positive association between religious identity and/or valuing religion with self-esteem (Smith, Weigert, & Thomas, 1979). Further research is needed that specifically focuses on American Indian youth.

CURRENT STUDY

Given the recent finding that ethnic identity was positively and significantly associated with self-esteem until the addition of religious identity into statistical models (Evans et al., 2014) and given the unique experiences of Lumbee youth, the current study aimed to further explore the relationship between these three constructs among Lumbee youth. It was hypothesized that religious identity would mediate the relationship between ethnic identity and self-esteem.

METHOD

The current sample was drawn from the Rural Adaptation Project (RAP), a 5-year longitudinal panel study of more than 6,000 middle-school students from 28 public schools located in two rural, economically disadvantaged counties in North Carolina. The data for the current study were collected at the RAP baseline in spring 2011, in year 2 of the study (spring 2012), and in year 3 of the study (spring 2013), providing temporal order for mediation analyses. The study received IRB approval from a major research university in the

Southeastern United States. Before completing the online assessment, students read a statement and then electronically signed their assent to participate. In accordance with school district policies, one of the rural counties adopted the assessment as part of normal procedures and all students were included on the study roster. The second county was geographically larger with a more extensive student population. Therefore, a random sample of 40% of middle school students was selected from district rosters. Parents in the second county received a letter explaining the study. If they did not want their child(ren) to participate, they returned the letter requesting non-participation and their child(ren) were removed from the study roster. Participants completed the online assessment in school computer labs under close supervision from research staff; to maintain confidentiality, surveys were identified using students' identification numbers rather than names.

Participants

The total sample over the first three years of the RAP project consisted of 6,790 youth in grades six through ten. However, for the current analysis, only data from American Indian participants were used. In addition, some participants were removed because their assessments were missing data for all variables of interest. Thus, the analyzed sample consisted of 1,571American Indian youth. About half (50.66%) of the sample were female and two thirds (76.11%) of the sample received free or reduced price lunch across all three years. In year three, the mean age was 14.22 years (*SD* = 1.49).

Measures

Data were obtained using a modified version of the School Success Profile (SSP; Bowen & Richman, 2008). The SSP is a 220-item youth self-report survey that measures attitudes and perceptions about school, friends, family, neighborhood, self, health, and well-being; this measure has been widely used and has well-documented reliability and validity (Bowen & Richman, 2008). The modified version, the School Success Profile Plus (SSP+), included 152 SSP items and three additional subscales: a modified version of the Rosenberg self-esteem scale (Rosenberg, 1965) and Phinney's Multigroup Ethnic Identity measure (Phinney & Ong, 2007).

Ethnic identity. The six-item version of Phinney's Multigroup Ethnic Identity Measure (MEIM: Phinney & Ong, 2007) was used to measure the strength of participants' ethnic identities. Example items included, "I have spent time trying to find out more about my ethnic group, such as its history, traditions, and customs" and "I have a strong sense of belonging to my own ethnic group." Each item was rated on a 5-point Likert scale (*Strongly Disagree, Disagree, Neither Agree Nor Disagree, Agree,* and *Strongly Agree*). Ethnic identity scores from year one were used in the current analysis and the Cronbach's alpha was .92 in this sample.

Religious identity. The three-item religious identity scale assessed the importance of religion in participants' lives. Example items included, "Religion plays an important role in my daily life," and "My religious faith influences the decisions I make." Each item was rated on a 3-point Likert scale (*Not Like Me, A Little Like Me,* and *A Lot Like Me*). Religious identity scores from year two were used in the current analysis and the Cronbach's alpha was .91 in this sample.

Self-esteem. In order to limit the length of the assessment, self-esteem was assessed with a five-item adapted version of the Rosenberg Self-Esteem scale (Rosenberg, 1965). The five items drawn from the Rosenberg scale were reworded as appropriate for the literacy level of a middle-school population. For example, superfluous and confusing words were removed from items to make them more easily understood. The item: "On the whole I am satisfied with myself" was re-worded to read: "I am satisfied with myself." Example items included, "I feel good about myself" and "I am able to do things as well as most other people." Each item was rated on a 3-point scale of *Not Like Me*, *A Little Like Me*, or *A Lot Like Me*. Self-esteem scores from year three were used in the current analysis and the Cronbach's alpha was .92 in this sample.

Analysis

Structural equation modeling (SEM) allows for the exploration of the relationships between latent variables as well as between observed and latent variables (Bowen & Guo, 2012). SEM is unique as it combines multiple regression equations and factor analysis (Hoyle, 2012) and permits researchers to test equations simultaneously, which is ideal for mediational analyses. When using SEM, it is preferable to have three waves of data (Cole & Maxwell, 2003) and the longitudinal nature of the current data was therefore well-suited for SEM. Before analyzing the data, cut off points for fit statistics (i.e., Comparative Fit Index [CFI], Tucker-Lewis Index [TLI], Root Mean Squared Error of Approximation [RMSEA]) were established using fit statistic criteria (West, Taylor, & Wu, 2012). A CFI and TLI greater than .95 and an RMSEA below .06 were used to assess adequate model fit. Although a non-significant chi square is desirable, this is quite rare, especially with a large sample size; thus, additional fit indices were used as well. Data were analyzed using MPlus and Mean- and Variance-adjusted Weighted Least Square (WLSMV) was used as an estimator. Indirect effects were calculated using the IND command in MPlus.

RESULTS

The model fit the data exceptionally well based on fit statistic criteria established by West and colleagues (2012). The chi-square value was 331.02(74), $p < .001$. The RMSEA value was .047, with a 90% confidence interval between .042 and .052. The CLI and TLI values were .995 and .994, respectively. The path from ethnic identity to religious identity, the path from ethnic identity to self-esteem, and the path from religious identity to self-esteem were significant. See Figure 1 for complete SEM model and path coefficients. The test for indirect effects was also significant ($p < .001$).

DISCUSSION

We hypothesized that religious identity would mediate the relationship between ethnic identity and self-esteem in a sample of Lumbee youth. This hypothesis was confirmed by the data analysis. The path from Time 1 ethnic identity to Time 3 self-esteem two years later was significant. This confirms the positive association between ethnic identity and self-esteem that has been found in previous research among several different racial/ethnic groups (e.g., Blash & Unger, 1995; Corenblum & Armstrong, 2012; Phinney & Chavira, 1992; Phinney et al., 2001; Umaña-Taylor & Updegraff, 2007). The salience of this relationship for Lumbee youth extends a previous study that reported a significant relationship between ethnic identity and self-esteem among Navajo youth (Jones & Galiher, 2007).

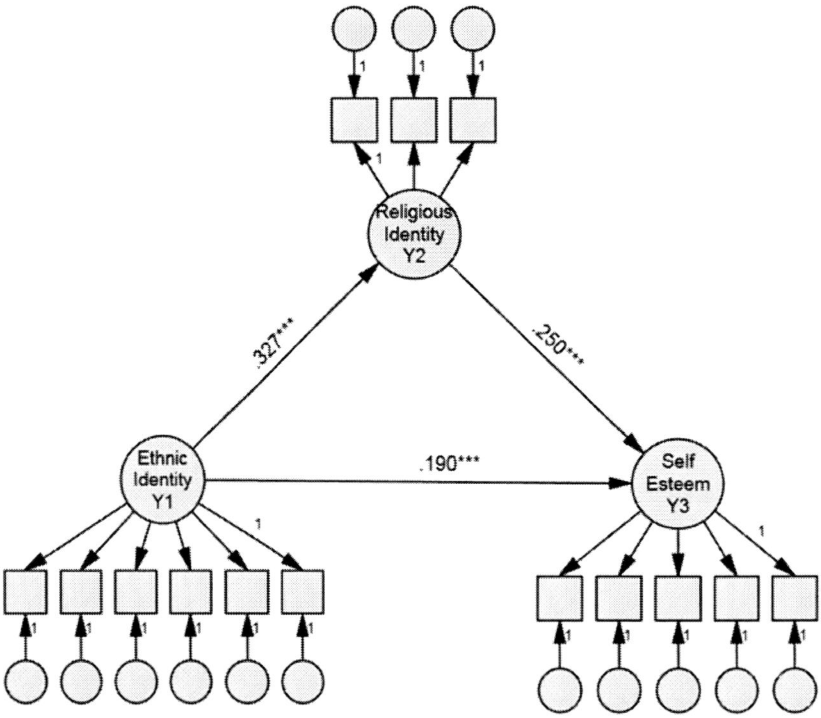

Note: Path coefficients are unstandardized. *** indicates p < .001. Lambda coefficients and correlations among endogenous variables are absent from the figure.

Figure 1. Structural Equation Model: Final Analysis Diagram.

Further, the longitudinal nature of our data shows that ethnic identity is connected to self-esteem two years later. Few prior studies have demonstrated the temporal order and longitudinal relationship needed to confirm the developmental link between ethnic identity and self-esteem for adolescents. In past cross-sectional studies, researchers have debated if high self-esteem fuels the investigation of ethnic identity or if ethnic identity increases positive feelings about oneself (Phinney &Chavira, 1992). Our longitudinal model provides evidence for the latter explanation, at least among Lumbee youth. While this is an important finding, caution is always warranted to making causal assertions because of unmeasured variables that may confound the results. Future studies should further explore the temporal order.

The current study also went beyond this foundational link to consider mediation through religious identity. The relationship between ethnic identity at Time 1 and religious identity one year later was significant. As predicted by social identity theory, ethnic and religious identities are often interconnected with changes in one identity often prompting changes in the other (Lopez, et al., 2011). Considering the importance that religion plays in the lives of Lumbee youth, we found this relationship between baseline ethnic identity and religious identity one year later to be very strong for our sample. The temporal order in our analyses provides a logical basis to believe that Lumbee youth who are investing in their ethnic identities may seek out churches with other Lumbee peers and mentors (see also, Dougherty, 2003; Markstrom, 1999).

Lumbee youth had both a direct path from ethnic identity to self-esteem and an indirect one through religious identity. Their self-esteem was related directly to both ethnic identity and religious identity. This finding also makes sense for a group that has both strong racial socialization for youth and close affiliations with religious institutions. Many American Indians (including the Lumbee) do not solely express their ethnic identities through religion. American Indians also hold Powwows, cultural awareness programs, storytelling, and mentoring from tribal elders (Garroutte et al., 2009). At the same time, religious identity is an important part of American Indians' lives without overshadowing their ethnic identities. Indeed, religious and cultural rituals and culture-based spirituality may be infused throughout everyday life for American Indians, keeping both pathways equally salient. Consequently, ethnic identity may influence self-esteem directly or through religious identity.

LIMITATIONS

The results of this study must be understood in light of the limitations. First, while the Lumbee sample provides insight into an understudied population, this unique sample limits the generalizability of the findings to other racial/ethnic groups as well as to other American Indian groups. Second, although every precaution was taken to ensure that filling out the assessments was a confidential process, youth might have been affected by the presence of their peers and therefore not responded to the survey questions honestly. Third, although Phinney's six-item MEIM (1992) has been empirically tested and used widely, ethnic identity is a complex construct and the six-item scale might not capture all aspects. Although a longer assessment of ethnic identity would have been ideal, this was not possible given time constraints for filling out the assessment.

CONCLUSION

Previous research has linked ethnic identity to self-esteem; however, few studies have focused on American Indian youth and even fewer studies have used longitudinal data to examine mediational mechanisms that might account for this relationship. This study investigated one such mediator, religious identity, which partially accounted for the relationship between ethnic identity and self-esteem in Lumbee youth. For these youth, ethnic identity and religious identity were equally important in building self-esteem. Using similar methodology, future research should more fully consider the complex ways in which different ethnic and racial groups balance culture and religion in identity development.

REFERENCES

Bagley, C. & Mallick, K. (1997). Self-esteem and religiosity: Comparison of 13- to 15-year-old students in Catholic and public junior high schools. *Canadian Journal of Education 22(*1), 89-92. doi:10.2307/1585814.

Bankston, C. L., & Zhou, M. (1995). Religious identity, ethnic identification, and adaptation of Vietnamese adolescents in an immigrant community. *Sociological Quarterly, 36*(3), 523-534. doi:10.1111/j.1533-8525.1995.tb00451.x.

Bernal, M. E., & Knight, G. P. (1993). *Ethnic identity: Formation and transmission among Hispanics and other minorities*. Albany, NY: State University of New York Press.

Blash, R. R., & Unger, D. G. (1995). Self-concept of African-American male youth: Self-esteem and ethnic identity. *Journal of Child and Family Studies, 4*, 359-373. doi:10.1007/BF02233968

Bourhis, R. Y. & Hill, P. (1982). Intergroup perceptions in British higher education: A filed study. In H. Tajefel (ed.), *Social identity and intergroup relations*. Cambridge University Press: Cambridge, UK.

Bowen, N., & Guo, S. (2012). *Structural equation modeling*. NY, NY: Oxford University Press.

Bowen, G. L., & Richman, J. M. (2008). *The School Success Profile*. Chapel Hill, NC: University of North Carolina. Retrieved from http://www.uncssp.org/

Bryant, A., & LaFromboise, T. D. (2005). The racial identity and cultural orientation of Lumbee American Indian high school students. *Cultural Diversity and Ethnic Minority Psychology, 11*, 82-89. doi:10.1037/1099-9809.11.1.82.

Chong, K. (1998). What it means to be Christian: The role of religion in the construction of ethnic identity and boundary among second-generation Korean Americans. *Sociology of Religion, 59,* 259-286. doi:10.2307/3711911.

Cole, D. A., & Maxwell, S.E. (2003). Testing mediational models with longitudinal data: Questions and tips in the use of structural equation modeling. *Journal of Abnormal Psychology, 112,* 538-577. doi:10.1037/0021-843X.112.4.558.

Corenblum, B., & Armstrong, H. D. (2012). Racial-ethnic identity development in children in a racial-ethnic minority group. *Canadian Journal of Behavioral Science, 44*(2), 124-137. doi:10.1037/a0027154.

Dial, A., & Eliades, D. K. (1975). *The only land I know: A history of the Lumbee Indians.* San Francisco: Indian Historian Press.

Dougherty, K.D. (2003). How monochromatic is church membership? Racial-ethnic diversity in religious community. *Sociology of Religion, 64*(1), 65-85. doi:10.2307/3712269.

Erikson, E. H. (1950). *Childhood and society.* New York, NY: Norton.

Evans, C. B. R., Smokowski, P. R., & Cotter, K. L. (2014). Individual characteristics, microsystem factors, and proximal relationship processes associated with ethnic identity in rural youth. *Journal of the Society for Social Work and Research, 5*, 45-77. doi:10.1086/675848.

Garroutte, E.M., Beals, J., Keane, E.M., Kaufman, C., Spicer, P., Henderson, J...& The AI-SUPERPFP Team. (2009). Religiosity and spiritual engagement in two American Indian populations. *Journal for the Scientific Study of Religion, 48*(3), 480-500. doi:10.1111/j.1468-5906.2009.01461.x.

Hoyle, R. H. (2012). Introduction and overview. In R. H. Hoyle (Ed.), *Handbook of structural equation modeling* (pp. 3-16). New York, NY: Guilford Press.

Jones, M. D. & Galliher, R. V. (2007). Ethnic identity and psychosocial functioning in Navajo adolescents. *Journal of Research on Adolescence, 17,* 683-696. doi:10.1111/j.1532-7795.2007.00541.x.

Keyes, C.L.M & Reitzes, D.C. (2007). The role of religious identity in the mental health of older working and retired adults. *Aging & Mental Health, 11*(4), 434-443. doi:10.1080/13607860601086371.

Le, T.N., Tov, W., & Taylor, J. (2007). Religiousness and depressive symptoms in five ethnic adolescent groups. *International Journal for the Psychology of Religion, 17*(3), 209-232. doi:10.1080/105086 107014 02259.

Lopez, A. B., Huynh, V. W., & Fuligni, A. J. (2011). A longitudinal study of religious identity and participation during adolescence. *Child Development, 82*, 1297-1309. doi:10.1111/j.1467-8624.2011.01609.x.

Lowery, M. M. (2010). *Lumbee Indians in the Jim Crow south*. Chapel Hill, NC: University of North Carolina Press.

Martinez, R.O. & Dukes, R.L. (297). The effects of ethnic identity, ethnicity, and gender on adolescent well-being. *Journal of Youth and Adolescence, 26*(5), 503-516. doi:10.1023/A:1024525821078.

Markstrom, C. A. (1999). Religious involvement and adolescent psychosocial development. *Journal of Adolescence, 22,* 205-221. doi:10.1006/ jado.1999.0211.

Nishina, A., Bellmore, A., Witkow, M. R., & Nylund-Gibson, K. (2010). Longitudinal consistency of adolescent ethnic identification across varying school ethnic contexts. *Developmental Psychology, 46*, 1389-1401. doi:10.1037/a0020728.

Phinney, J. S. (1990). Ethnic identity in adolescents and adults: Review of research. *Psychological Bulletin, 108,* 499-514. doi:10.1037//0033-2909.108.3.499.

Phinney, J. S., & Chavira, V. (1992). Ethnic identity and self-esteem: An exploratory longitudinal study. *Journal of Adolescence, 15,* 271-281. doi:10.1016/01401971(92)90030-9.

Phinney, J. S., Cantu, C. L., & Kurtz, D. A. (1997). Ethnic and American identity as predictors of self-esteem among African American, Latino, and White adolescents. *Journal of Youth and Adolescence, 26,* 165-185. doi:10.1023/A:1024500514834.

Phinney, J. S., Horenczyk, G., Liebkind, K., & Vedder, P. (2001). Ethnic identity, immigration, and well-being: An interactional perspective. *Journal of Social Issues, 57*, 493-510. doi:10.1111/0022-4537.00225.

Phinney, J. S. & Ong, A. D., (2007). Conceptualization and measurement of ethnic identity: Current status and future directions. *Journal of Counseling Psychology, 54,* 271-281. doi:10.1037/0022-0167.54.3.271.

Pittenger, S.M. (1998). The relationship between ethnic identity, self-esteem, emotional well-being and depression among Lakota/Dakota Sioux adolescents. *ProQuest Dissertations and Theses Full Text.*

Rosenberg, M. (1965). *Society and the adolescent self-image.* Princeton, NJ: Princeton Univ. Press.

Smith, C.B., Weigert, A.J., & Thomas, D.L. (1979). Self-esteem and religiosity: An analysis of Catholic adolescents from five cultures.*Journal for the Scientific Study of Religion, 18*(1), 51-60. Retrieved from http://www.jstor.org/stable/ 1385378.

Stannard, D. E. (1992). *American Holocaust: The conquest of the new world.* New York, NY: Oxford University Press.

Tajfel, H., & Turner, J. (1979). An integrative theory of intergroup conflict. In W Austin & S. Worchel (Eds.), *The social psychology of intergroup relations* (pp. 33-47). Monterey, CA: Brooks/Cole.

Tajfel, H. (1982). *Social identity and intergroup Relations.* New York, NY: Cambridge University Press.

Umaña-Taylor, A. J., & Updegraff, K. A. (2007). Latino adolescents' mental health: Exploring the interrelations among discrimination, ethnic identity, cultural orientation, self-esteem, and depressive symptoms. *Journal of Adolescence, 30*, 549-567. doi:10.1016/j.adolescence.2006.08.002.

West, S. G., Taylor, A. B., & Wu., W. (2012). Model fit and model selection in SEM. In R. A. Hoyle (Ed.), *Handbook of structural equation modeling* (pp. 209-231). New York, NY: Guilford Press.

In: Native Americans
Editor: Jerome Mendoza

ISBN: 978-1-63483-184-0

Chapter 4

THE EFFECT OF FOOD ASSISTANCE ON AMERICAN INDIAN WOMEN'S FOOD CHOICES

Ariun Ishdorj*[*] and Victoria Salin
Department of Agricultural Economics,
Texas A&M University, TX, US

ABSTRACT

Limited access to nutrient-rich foods in low-income neighborhoods as well as limited income resources may collectively result in poor diet quality. Federal food and nutrition assistance programs support one in four Americans in achieving adequate nutrition. The Special Supplemental Nutrition Program for Women, Infants, and Children (WIC) is one of the food assistance programs that plays an important role in improving food choices and diet quality in vulnerable populations due to its targeted impact and broad reach.

The WIC program serves participants by providing supplemental foods, nutrition education, breastfeeding support and referrals to health and social services. The goals are to improve birth outcomes, to support the growth and the development of infants and children, and the promotion of long-term health of all WIC participants. In 2009, the U.S. Department of Agriculture (USDA) instituted change in the types and the

[*] Corresponding Author: Ariun Ishdorj, aishdorj@tamu.edu.

amount of foods offered to WIC participants. One of the main changes in WIC food packages includes the provision of vouchers for purchases of any eligible fresh, frozen, or canned fruits and vegetables. The revised WIC food packages were intended to improve the diets and health of culturally diverse WIC participants with a wide range of traditional food preferences. American Indians are a subgroup of the population that is served by the WIC program. Compared with the general U.S. population, the WIC population is distinctively poor, but American Indian WIC participants face additional risks as nearly 81.2% are at or below 130% of the poverty. Although it is documented that obesity, diabetes, heart disease and cancer are major problems affecting American Indians relatively little is known about their eating behaviors. Critical factors present within Indian reservations may limit the ability to access and consume fruits and vegetable as well as other healthy foods. This study addresses the effect of WIC food-package revisions on frequency of juices and frequency and variety of fruits and vegetables consumed by American Indian WIC women. There were modest changes in food choice after the revisions in the food package. Specifically, the consumption of vegetables and some fruits increased, while the consumption of juice declined in frequency.

INTRODUCTION

Many of the diseases that plague our modern society come in large part from poor food choices which contribute to the rise of major chronic diseases, including overweight and obesity. The Dietary Guidelines for Americans (DGA) constitute the most important piece of nutrition advice in Federal food, education and information programs in the United States (Davis and Saltos 1999). The last release of DGA was in 2010 (DGA 2010), with a major objective to increase the consumption of fruits and vegetables and other healthy foods.

Fruits and vegetables provide key nutrients and dietary fiber which have protective effect against many chronic diseases. Increased consumption of fruits and vegetables has been shown to promote heart health, reduce risk of certain cancer, improve diet quality, and help individuals to manage their body weight. However, very few changes in consumption have been observed as a result of DGA. In fact, consumption of healthy foods is well below the recommended levels, while consumption of calorie-dense empty-nutrient food is above the recommendation. The typical American diet only meets 42% of the fruit recommendations, 59% of vegetables, 15% whole grains, 52% dairy,

44% seafood and 61% oils. On the other hand, meat, poultry and eggs exceed the recommended levels with 110% of the recommendation; calories from solid fats and added sugars are 280% of the recommendation; refined grains are 200% of the recommendation; and saturated fat 110% of the recommendation (DGA 2010).

While average Americans are not meeting the recommended dietary guidelines, people who live in low-income households are even less likely to meet recommendations for fruits, vegetables, and fiber than higher income consumers (Satia 2009; Casagrande and et al. 2007). For example, an analysis of the National Health and Nutrition Examination Survey 2003-2004 data (USDHHS 2005) shows that adults who live in lower-income households are less likely to consume any fruits and vegetables, and have fewer people who meet minimum recommendations for adults. Nutritious foods, such as fruits and vegetables, may not be affordable to all individuals, especially to those who are low-income and below poverty.

Improving nutritional well-being of low-income population is a primary objective of U.S. food assistance programs. The third largest of those is the Special Supplemental Nutrition Assistance Program for Women, Infants, and Children (WIC). WIC serves over half of all infants born in the United States, about one-quarter of all U.S. children of ages 1 through 5 years, along with many of their mothers (USDA/FNS 2011; IOM 2006). The WIC program provides supplemental foods, nutrition education, breastfeeding support and referrals to health and social services for its participants. The WIC program was established in 1972, on the basis that pregnant women, infants, and young children are at risk to their physical and mental health due to poor or inadequate nutrition (Oliveria et al. 2002). To expand the selection of foods the program offered and help combat the problem of obesity WIC food packages were revised in 2009. The new packages were aligned with the 2005 Dietary Guidelines for Americans and infant feeding practices supported by the American Academy of Pediatrics, and were intended to better support the establishment of successful, long-term breastfeeding; provide WIC participants with a wider variety of healthy foods; and allow state agencies greater flexibility in prescribing packages that accommodate cultural preferences of participants. To implement the various food packages, most WIC participants receive vouchers to use in purchasing supplemental foods (including fruits and vegetables) at authorized retail outlets. The analysis of the impact of these changes to the vouchers is limited by the problem of inferring what fruit and vegetable purchases/consumption would have been in absence of the program. For example, the voucher-purchased foods may merely substitute for fruits and vegetables that would have been purchased in the absence of the program.

Cole, et al. (2011) found that 19 states allowed participants to use multiple vouchers in a single transaction, and 38 states allowed participants to pay the difference on the cost of fruits and vegetables beyond the value of the vouchers. However, Gleason and Pooler (2011) found that in Wisconsin there was confusion among supermarkets, checkout clerks, and participants over buying more fruits and vegetables than the value of the voucher. While most participants (93%) initially used the supplemental vouchers, full use of the vouchers' value was found to be only 56.9% among Hispanics. However, more than 63% of WIC families used their fruit and vegetable vouchers to purchase more fruits and vegetables than the maximum value of the vouchers, which may have been even higher were it not for policy confusion regarding such purchases.

Limited availability and accessibility to affordable and nutritious food in low-income communities and neighborhoods can result in poor diet quality (Bitler and Haider 2011; Larson, Story, and Nelson 2009). It is clear that among the segments of the population which are in the greatest need to improve nutrition are those in poverty, who lack ready access to nutritious local food, who live in areas sometimes referred to as "food deserts." Food desert defines as an "area in the United States with limited access to affordable and nutritious food, particularly such an area composed of predominantly lower income neighborhoods and communities…" (VerPloeg and Breneman 2013; Dutko et al. 2012). Many in these population segments receive food assistance benefit, although some eligible individuals and households choose not to participate in food assistance programs. The "food desert" concept emphasizes the interactions among food access/availability, poverty, and nutrition (VerPloeg and Breneman 2013; VerPloeg e al. 2009; VerPloeg 2010a; VerPloeg 2010b; Dutko 2012). These interactions have been studied in both urban and rural areas (Weatherspoon et al. 2012; Kyureghian et al. 2013; Alviola et al. 2013), and emphasize the lack of availability of supermarkets, superstores, or large grocery stores as key factors influencing nutrition and food choices (Rose et al. 2009; Alviola et al. 2013). The food desert research amply demonstrates that in underserved and poor communities, the result of limited access to healthy food is a poor diet and consequent adverse health effects (Bitler and Haider 2011; Larson et al. 2009). However, the comprehensive reviews of studies of the impacts of store location and distance on fruit and vegetable consumption have led to inconsistent conclusions (Beaulac et al. 2009). For example, Blanchard and Lyson (2002) concluded that people who were located in rural areas, with longer travel distances for food access, were less like to consume the recommended level of fruits and vegetables. Dong and Lin (2009) estimated that a 10% price subsidy would encourage low-income Americans to increase their consumption of fruits by 2.1-5.2% and vegetables by 2.1-4.9%. Even so, most low-income Americans still would not meet Dietary Guideline recommendations. As implied by Dong and Lin (2009), the impact of a subsidy, coupon, or cash voucher program is influenced by the baseline level of fruit and vegetable purchases. Hence, limited access to nutrition rich foods in low-income neighborhoods as well as limited income resources may or may not collectively influence these individuals fruits and vegetables consumption.

Inadequate consumption of fruits and vegetables has been documented among individuals living in low-income rural households and those of minority race/ethnicity. American Indians belong to this group. Majority of Indian Tribal Organizations (ITO) are located in the areas designated as food deserts and, generally, have fewer resources. Poverty rates among American Indians are high (U.S. Census Bureau 2004). American Indians were historically at risk for malnourishment due to their poverty status (U.S. Department of Agriculture 2002). Poverty also is associated with inaccessibility to food (O'Connell, Buchwald and Duncan 2011; Casey, et al. 2001), and poor dietary habits such as the consumption of less healthy, high-energy density foods (Drewnowski and Specter 2004). For American Indians, the combination of these issues may contribute to their eating behaviors and subsequent health status among this population. However, we know less about the current demographic and health situation of American Indians than that of other racial or ethnic group (Sandefur, Rindfuss and Cohen 1996) due to data limitations.

In 2010, Native American enrollment in the WIC Program was 10.5% (U.S. Department of Agriculture 2011) which is highly disproportionate to their representation in the U.S. population (1.7%) (Norris, Vines and Hoeffel 2012). Compared with the general U.S. population, the WIC population is distinctively poor, with about 69.6% of WIC participants at or below the poverty line, compared with 14.3% of the general population (U.S. Department of Agriculture 2011). Native American WIC Participants face additional risks as nearly 81.2% are at or below 130% of the poverty level (U.S. Department of Agriculture 2011). While the situation reveals high risk among this group for nutrition related health conditions, it also highlights prevention and intervention opportunities.

WIC benefits are administered by each state. However, State ITOs are considered sovereign entities, and are run independently of the geographic state in which they are located(U.S. Department of Agriculture 2002). Although many studies looked at the effect of revised WIC food packages on program participants and availability and accessibility of foods through WIC vendors, to our knowledge, there has been little research (Ishdorj and Capps 2013) evaluating the changes in the WIC on American Indians WIC participants food intake. Therefore, a clear need exists for information on nutritional behaviors of individuals on *and* off ITOs.

One of the main changes in WIC food packages includes the addition of vouchers for fruits and vegetables to the packages of WIC women and children of ages two through four. Given this important change in the delivery of the program, it is of interest to know whether the provision of vouchers for fruits and vegetables has in fact helped to increase consumption of fruits and vegetables. Therefore, the overall goal of this paper is to better understand the food preferences and consumption behaviors of American Indian women participating in WIC. Our specific objective is to explore the type of fruits and vegetables consumed and the frequency of consumption of juices, fruits, and vegetables on the part of American Indian WIC women. In addition, this study will enable examination of before and after the changes resulting from the WIC new food package and the comparison of outcomes for American Indian WIC women residing on and off ITOs.

THE MODEL

The most fundamental proposed change to the WIC food packages is the provision of vouchers in the food packages for all WIC participants six months of age and older to purchase a wide variety of fresh fruits and vegetables. In special cases, when fresh fruits and vegetables are of limited availability, canned, dried or frozen fruits and vegetables are allowed.

In our data the consumption of juices, fruits, and vegetables by WIC women is expressed as frequency of consumption, such as:

> 'never'; '1-3 times per week'; '4-7 times per week'; and '2 or more times per day'.

The respondent variables are expressed as an ordinal ranking, and consequently, an ordered probit model was used to assess the effect of socioeconomic characteristics on the frequency of juice, fruits, and vegetables consumed by American Indian WIC women on and off ITOs.

The ordered probit model is specified around the latent variable whose level is affected by explanatory variables such that:

$$y^* = \beta' X + \varepsilon, \epsilon \sim N(0,1)$$

where X is a matrix of independent variables, y* is unobserved latent variable, β' is a vector of parameters associated with X, and ϵ is a vector of stochastic error terms that is not captured by the design matrix X. The relationship between y and y*, where y is observed, is assumed to be a function of cut-off points (μ_j) which are estimated along with the regression coefficients and vary with observations. Assume an individual's frequency of consumption is a choice of one of the four frequencies (y = 1 never; y = 2; y = 3; and y = 4) corresponding to the Likert scale such that:

$$\begin{aligned} &y = 0 \text{ if } y^* \leq 0 \\ &y = 1 \text{ if } 0 < y^* \leq \mu_1 \\ &y = 2 \text{ if } \mu_1 < y^* \leq \mu_2 \\ &y = 3 \text{ if } \mu_3 \leq y^* \end{aligned}$$

where μ_1, and μ_2 are unknown threshold parameters of y* to be estimated. The interpretation of the estimated coefficients requires the calculation of the category specific marginal effects since the changes in the probabilities associated with the intermediate categories (1 through j-1) cannot be signed a priori. The marginal effects should sum to zero by cancelling one another out across the response categories.

DATA

To address the aforementioned issues, we used data from the National Food and Nutrition Survey for WIC (NATFAN). NATFAN consisted of two repeated cross-sectional surveys of WIC participants from multiple states, administered before and after revisions to the WIC program. For this paper, we used results from NATFAN surveys of WIC women, collected by state and ITO WIC programs before and after changes in WIC food packages. During 2009, 40 states and 11 ITOs participated in administering questionnaires in WIC clinics before the food package revisions. The survey was readministered at least six months following the food package changes, late in 2010 and early 2011. Individual state programs administered surveys using samples of WIC participants who attended WIC clinics during the data collection periods. The questionnaire focused on WIC participants demographics with relations to preferences, feeding and consumption practices. In addition to socio-demographic information, response items captured the frequency and type of fruits and vegetable consumption.

Our observations correspond to an independently pooled cross-section dataset. Since observations were randomly drawn both before and after the rollout of the new WIC food packages, the pooled sample also is random and ordered probit model was estimated using the pooled data.

RESULTS

Descriptive statistics of our sample for before and after the changes in WIC food packages are reported in Table 1. For descriptive analysis, we divided our sample WIC women into three subgroups, American Indians on ITOs, other race on ITOs, and American Indians off ITOs. These subgroups allow us to make several comparisons of juice, fruit and vegetable intakes between American Indian and non-American Indian WIC women on ITOs and between American Indian WIC women on and off ITOs, both before and after the changes in WIC food packages. The majority of American Indians in our sample lived off ITOs. This is consistent with current statistics. In 2010, the majority of the American Indian population (78%) lived outside of the Indian Tribal Organizations (ITOs) (Norris, Vines and Hoeffel 2012).

The three categories of WIC women: 'pregnant,' 'postpartum' and 'currently breastfeeding' are not mutually exclusive. It is highly likely that women who are postpartum are also breastfeeding. About 49% of women in our sample were pregnant, 4% currently breastfeeding and 28% had a baby within last six months. Slight more than 1% of our sample respondents reported being pregnant and currently breastfeeding and 16% reported currently breastfeeding and had a baby within the last six month. The average age of WIC participating women in our sample was 24 years. Compared to non-American Indian WIC women on ITOs, American Indian WIC participants, both on and off ITOs, were relatively younger and less likely to have college or graduate degrees.

We observed a significant increase in the percentage of women reported 'never' consuming fruit juice after the changes in WIC (Table 2). Since amount of juice provided by WIC in women and children packages was reduced, these findings indicate that the change caused a shift in the intended direction with respect to juice consumption. Significant decreases in the frequency of vegetables consumed ‘never’ and ‘1-3 times per week’ and increase in the frequency of vegetable consumed ‘2 or more times per day’were observed for American Indian women off ITOs. We did not observe significant changes in the frequency of fruits and frequency of vegetables consumed for WIC women on ITOs, both American Indian and non-American Indian.

Fruits and vegetables vary considerably in quality, and price and these can have significant impacts on individual’s quantity and variety of food consumed. While the surveys analyzed in this study did not contain information on price and quality attributes, it was possible to assess the respondents’ general preferences for specific types of fruits and vegetables consumed. The survey question asked respondents to list fruits and vegetables that were consumed and a rank was created by summation of the number of respondents who had ever consumed that specific food item. The rankings were compared for before and after the changes in WIC (Tables 3 and 4). We did not observe significant differences in the variety of fruits and variety of vegetables consumed by American Indian women on ITOs, non-American Indian women on ITOs and American Indian women off ITOs. Hence, we only reported rankings of fruits, not including juices, and vegetables for the entire sample before and after changes in WIC. With respect to vegetables, corn and potatoes were the most popular vegetables consumed by WIC women both before and after the changes in WIC. We observed significant increases in the percentage of some vegetables consumed after the changes in WIC. Specifically, significantly higher percentages of WIC participants reported consuming lettuce, carrots, broccoli, green beans, onions, avocados, spinach, squash, okra, greens and sprouts after the issuance of vouchers for fruits and vegetables. With respect to fruits, apples and bananas were the most popular fruits consumed by WIC women both before and after the changes in WIC. Significantly higher percentage of WIC women reported consuming grapes, melons, and cherries after the changes in WIC food packages.

The estimated marginal effect of ordered probit model for frequency of juice, fruits, and vegetables consumed are reported in Tables 5, 6, and 7, respectively. In order to capture the effect of changes in WIC food packages on the frequency of juices, fruits and vegetable consumed we included a survey-period dummy variable (labeled as *post*).

Table 1. Demographic Characteristics American Indian Women in WIC

	On ITO American Indian		On ITO Other Race		Off ITO American Indian		Total	
	Before	After	Before	After	Before	After	Before	After
Pregnant	49.0	52.1	42.0	52.1	53.8	56.1	52.1	54.7
Breastfeeding	26.9	18.1	27.2	17.0	20.2	22.3	21.7	20.7
Postpartum	47.6	48.3	61.1	49.1	45.0	40.7	46.7	43.5
Age								
Under 18 years	12.4	11.3	8.0	8.5	18.4	16.3	16.7	14.2
19-24 years	41.4	42.0	43.1	42.4	45.2	44.6	44.4	43.7
25-29 years	24.8	23.9	26.0	26.7	21.1	23.5	22.1	24.0
30-34 years	13.8	18.5	17.0	14.5	10.2	9.1	11.3	11.9
35 or more years	7.6	4.3	5.9	7.9	5.1	6.5	5.5	6.2
Education								
Less than high school	33.8	26.0	30.1	26.1	38.9	34.5	37.4	31.5
High school or GED	31.9	37.0	34.9	38.2	31.6	33.8	31.9	35.1
Some college	31.7	34.5	29.0	31.5	26.6	29.4	27.5	30.8
College or graduate	2.6	2.5	6.0	4.2	2.9	2.3	3.2	2.6
Language most often spoken at home								
English	94.5	95.0	63.0	77.6	69.4	67.8	72.2	75.0
Spanish	3.4	2.1	23.8	10.3	20.0	19.7	18.2	14.6
Both English and Spanish	0.0	0.4	12.3	11.5	10.3	11.9	9.1	9.4
Other language	2.1	2.5	0.9	0.6	0.3	0.6	0.5	1.0
Region[a]								
Northeast	-	-	-	-	10.1	5.4	7.9	3.5

	On ITO American Indian		On ITO Other Race		Off ITO American Indian		Total	
	Before	After	Before	After	Before	After	Before	After
Mid-Atlantic	-	-	-	-	9.3	9.0	7.3	5.8
Southeast	-	-	-	-	9.5	5.0	7.4	3.6
Midwest	-	-	-	-	3.2	5.4	2.5	3.6
Southwest	-	-	-	-	24.4	23.0	30.0	37.9
Mountain Plains	-	-	-	-	26.0	32.8	23.3	25.5
Western	-	-	-	-	17.5	19.4	21.6	20.1
N	*145*	*238*	*100*	*165*	*857*	*725*	*1,102*	*1,128*

[a]Means for regions for ITO and non-ITO are not reported for confidentiality reasons.

Table 2. Frequency of Juice, Fruit, and Vegetable Consumption

	On ITO American Indian			On ITO Other Race			Off ITO American Indian			Total		
Frequency (%)	Before	After	P-value	Before	After	P-value	Before	After	P-value	Before	After	P-value
Juice												
Never	2.76	13.87	<0.001	5.1	12.12	0.03	6.65	8.41	0.16	5.99	10.11	<0.001
1-3 times/wk	34.48	38.24	0.45	32	35.76	0.53	27.65	29.1	0.27	28.95	32	0.1
4-7 times/wk	32.41	22.27	<0.001	29.9	31.52	0.79	31.04	30.62	0.68	31.13	28.99	0.27
2 or more times/d	30.34	25.63	0.35	33	20.61	0.02	34.66	31.86	0.18	33.94	28.9	0.01
Fruit												
Never	2.07	3.36	0.46	4.8	4.24	0.77	3.27	2.9	0.35	3.27	3.19	0.62
1-3 times/wk	21.38	23.11	0.51	31.2	16.36	<0.001	22.17	16.14	<0.001	22.87	17.64	0.002
4-7 times/wk	34.48	34.87	0.93	35.9	43.64	0.22	33.49	33.24	0.91	33.85	35.11	0.53
2 or more times/d	42.07	38.66	0.51	28.1	35.76	0.19	41.07	47.72	0.005	40.02	44.06	0.03

Table 2. (Continued)

	On ITO American Indian			On ITO Other Race			Off ITO American Indian			Total		
Frequency (%)	Before	After	P-value	Before	After	P-value	Before	After	P-value	Before	After	P-value
Vegetable												
Never	3.45	4.2	0.37	4.1	4.24	0.92	5.48	3.45	0.05	5.08	3.72	0.1
1-3 times/wk	24.83	27.73	0.45	27.9	19.39	0.1	24.27	20.55	0.07	24.68	21.9	0.9
4-7 times/wk	37.93	35.29	0.61	40	43.64	0.56	39.32	37.38	0.42	39.2	37.85	0.51
2 or more times/d	33.79	32.77	0.83	28	32.73	0.42	30.92	38.62	<0.01	31.03	36.52	<0.001
N	145	238		100	165		857	725		1,102	1,128	

Table 3. Ranking of Fruits Consumed, Before and After the Changes in WIC

Rank	Fruits	Before	After	p-value
1	Apples	90.02	90.25	
2	Bananas	89.20	89.98	
3	Oranges	85.57	84.84	
4	Grapes	84.66	87.68	0.03
5	Strawberries	82.30	86.88	
6	Watermelon	70.78	78.90	
7	Peaches	64.97	67.02	
8	Pineapple	61.80	65.51	
9	Melons	56.72	63.03	0.002
10	Pears	47.28	48.32	
11	Lemon or Limes	43.19	44.33	

Rank	Fruits	Before	After	p-value
12	Cherries	42.56	50.98	<0.001
13	Mangos	38.93	37.32	
14	Berries	34.39	36.97	
15	Kiwis	34.03	33.42	
16	Plums	28.49	29.26	
17	Grapefruit	26.23	21.19	0.05
18	Raisins	25.23	23.40	
19	Tangerines	23.77	21.99	
20	Nectarines	20.15	25.53	0.03
21	Papayas	12.16	11.70	
22	Fresh Apricots	10.62	10.73	
23	Prunes	7.44	6.38	
24	Dried Apricots	4.99	4.52	
25	Figs	4.81	2.93	0.02
26	Rhubarb	3.18	3.01	
27	Other	2.36	1.77	
28	Dates	1.81	1.60	0.02
	N	1,102	1,128	

Table 4. Ranking of Vegetables Consumed, Before and After the Changes in WIC

Rank	Vegetables	Before	After	p-value
1	Corn	89.11	89.80	
2	Potatoes	86.75	88.03	
3	Lettuce	81.03	86.26	<0.001
4	Carrots	79.22	80.14	0.01
5	Tomatoes	76.23	80.67	
6	Broccoli	73.05	76.06	0.1
7	Green Beans	70.24	73.85	0.05
8	Cucumber	67.06	68.79	
9	Onions	59.98	63.30	0.1
10	Peppers	49.73	52.84	
11	Green Peas	45.19	47.43	
12	Cabbage	43.74	46.10	
13	Avocados	43.10	47.16	0.05
14	Cauliflower	37.93	39.98	
15	Mushrooms	35.75	37.06	
16	Sweet Potatoes	31.13	28.37	
17	Spinach	28.04	32.27	0.02
18	Squash	23.32	28.28	<0.001
19	Asparagus	20.51	20.92	
20	Okra	15.25	17.82	0.1
21	Greens	14.25	17.38	0.04
22	Tomatillos	13.25	11.70	
23	Winter Squash	12.34	14.01	
24	Brussels Sprouts	8.98	11.97	0.02

25	Beets	7.17	8.60	
26	Eggplant	6.81	5.94	
27	Chayote	6.35	6.12	
28	Other	1.45	1.51	
	N	1,102	1,128	

Table 5. Marginal Effects of Ordered Probit Model for Juice Consumption

Variable		Never			1-3 times/week			4-7 times/week			2 or more/day	
Pregnant		-0.011			-0.017			0.002			0.025	
Postpartum		0.014			0.022			-0.003			-0.033	
Breastfeeding		-0.030	***		-0.047	***		0.006	***		0.072	***
Under 18 years		0.017			0.026			-0.003			-0.040	
19-24 years		0.006			0.010			-0.001			-0.015	
25-29 years		-0.008			-0.013			0.002			0.020	
30-34 years		0.033	**		0.051	**		-0.006	*		-0.077	**
Less than high school		-0.053	***		-0.083	***		0.010	**		0.126	***
High school or GED		-0.037	*		-0.057	*		0.007	*		0.087	*
Some college		-0.046	**		-0.072	**		0.008	**		0.109	**
ITO		0.023	***		0.036	***		-0.004	***		-0.054	***
Post survey		0.025	***		0.039	***		-0.005	***		-0.059	***

***,**, and * represent significance at 1%, 5% and 10% levels, respectively.

Table 6. Marginal Effects of Ordered Probit Model for Fruit Consumption

Variable	Never		1-3 times/week		4-7 times/week		2 or more/day	
Pregnant	-0.020	***	-0.065	***	-0.023	***	0.108	***
Postpartum	-0.002		-0.007		-0.003		0.012	
Breastfeeding	-0.022	***	-0.070	***	-0.025	***	0.117	***
Under 18 years	0.005		0.017		0.006		-0.028	
19-24 years	0.001		0.004		0.001		-0.007	
25-29 years	-0.005		-0.017		-0.006		0.028	
30-34 years	0.004		0.011		0.004		-0.019	
Less than high school	-0.017	*	-0.056	*	-0.020	*	0.093	*
High school or GED	-0.011		-0.036		-0.013		0.059	
Some college	-0.014		-0.047		-0.017		0.078	
ITO	0.011	***	0.036	***	0.013	***	-0.059	***
Post survey	-0.010	***	-0.032	***	-0.011	***	0.053	***

***,**, and * represent significance at 1%, 5% and 10% levels, respectively.

Table 7. Marginal Effects of Ordered Probit Model for Vegetable Consumption

Variable		Never			1-3 times/week			4-7 times/week			2 or more/day	
Pregnant		0.001			0.003			0.000			-0.004	
Postpartum		0.008			0.021			0.003			-0.032	
Breastfeeding		-0.027	***		-0.070	***		-0.009	***		0.106	***
Under 18 years		0.043	***		0.113	***		0.014	***		-0.171	***
19-24 years		0.020	**		0.052	**		0.006	**		-0.078	
25-29 years		0.007			0.018			0.002			-0.027	
30-34 years		0.011			0.030			0.004			-0.045	
Less than high school		0.016			0.042			0.005			-0.063	
High school or GED		0.021	*		0.055	*		0.007			-0.082	**
Some college		0.009			0.024			0.003			-0.036	
ITO		0.009	**		0.023	*		0.003	*		-0.035	*
Post survey		-0.013	***		-0.035	***		-0.004	***		0.052	***

***,**, and * represent significance at 1%, 5% and 10% levels, respectively.

The *post* dummy takes a value of zero (post = 0), if the survey was conducted before the changes in the WIC food packages and takes a value of one (post = 1),if the survey was conducted after food package changes. The estimated marginal effects allow us to directly compare the magnitude of the relationship between the explanatory variables in the model and the likelihood that a WIC participant reported frequency of 'never,' '1-3 times per week,' '4-7 times per week,' and '2 or more times per day'. We included a dummy variable ITO to capture if the lives on ITO territory.

Our findings from multivariate analysis are consistent with findings from descriptive analysis. The marginal effects of WIC women in the post survey indicate an increase in the likelihood of 'never' consume juice and '1-3 times per week' and decrease in the likelihood of consuming '4-7 times per week' and '2 or more times per day' (Table 5). Living on ITO is associated with increase in probability of consuming juice less frequently. Education level has a significant effect on the frequency of juice consumption. Compared to WIC women with college or graduate degree women with high school or some college education had higher probability of consuming juice more frequently. Compared to WIC women who are not currently breastfeeding, WIC women who are currently breastfeeding were 0.6% and 7.2% more likely to consume juice '4-7 times per week' and '2 or more times per day,' respectively.

With respect to frequency of fruit consumption, the marginal effect indicates that pregnant, and breastfeeding women were 10.8% and 11.7% more likely to consume fruits '2 or more times per day'. Frequency of consuming fruit '2 or more times per day' increased by 5.3% after the changes in WIC food packages. However, compared to WIC women off ITOs, WIC women on ITOs were 5.9% less likely to consume fruits '2 or more times per day'.

The marginal effects of explanatory variables on the frequency of vegetables consumed were similar to the results of fruits consumption. After the changes in WIC, we observed a decrease in the likelihood of consuming vegetables 'never,' '1-3 times per week' and '4-7 times per week' and increase in the likelihood of consuming vegetables '2 or more times per day'. WIC women on ITOs were 3.5% less likely to consume vegetables compared to WIC women off ITOs.

CONCLUSION

WIC is considered as a premier food assistance program and plays an important role in providing nutritious food and nutrition and breastfeeding education to its low-income target population. The objectives of this study were to assess the effect of changes in WIC food package on American Indian WIC women, a population subgroup for which there is a gap in the existing literature, and frequency of juice, fruit and vegetable consumption. This study provides new quantitative evidence with descriptive analysis and ordered probit models for a cross sectional data of WIC participants before and after changes in WIC and WIC food packages. We found that provision of vouchers for fruits and vegetables had an intended positive effect on frequency of fruits and vegetables consumption. Frequency of juice consumption had significantly decreased after changes in WIC, also an intended result of the program change. There were not many differences in choices overall, but some green vegetables were ranked higher after the voucher plan revision than had been the case in the earlier period.

These findings indicate that the food assistance program for pregnant women and mothers of young children had a limited effect, but the effect was in the direction of meeting the Dietary Guidelines. Like most Americans, the WIC participants in this study do not meet the goal of the Guidelines in consuming fruits and vegetables five times per day. However, after the vouchers were changed to cover more fruits and vegetables, there was some discernable increase in the frequency of consumption of those items. Women in the WIC program living on ITOs were less likely than those living off the ITO to fall into the category of consuming fruits and vegetables multiple times per day. However, WIC women who are American Indians reported consuming fruits and vegetables more frequently after the program coverage of fruits and vegetables was expanded. The impact of the assistance program was likely high for the subgroup of American Indians because of the high incidence of poverty among this population. Admittedly, this study has its limitations. Given the data limitations, we only looked at the effect of changes in WIC on WIC women food consumption on and off ITOs. Future research needs to address the effect of WIC participation on participants' food intake and compare findings to those who are eligible but do not participate in WIC. Participation in other assistance programs, household composition, and access to retail outlets can play an important role in WIC participants' food consumption preferences.

ACKNOWLEDGMENT

This research was funded through a Research Innovation and Development Grant in Economics (RIDGE) Center for Targeted Studies award from the U.S. Department of Agriculture's Economic Research Service, administered through the Southern Rural Development Center at Purdue University.

REFERENCES

Alviola, P. A., Nayga, R. M., &Thomsen, M.. (2013). Food Deserts and Childhood Obesity. *Applied Economic Perspectives and Policy,* vol. 35, pp. 106-124.

Beaulac, J., Kristjansson, E., & Cummins, S.. (2009*). A systematic review of food deserts,* 1966-2007, Preventing chronic disease, vol. 6.

Bitler, M., &Haider, S.J. (2011). An Economic View of Food Desserts in the United States. *Journal of Policy Analysis and Management,* 30(1):153-176.

Blanchard, T. &Lyson, T. (2002). Access to low cost groceries in nonmetropolitan counties: Large retailers and the creation of food deserts. Measuring Rural Diversity Conference Proceedings, November, pp. 21-22.

Casagrande, S.S., Wang, Y., Anderson, C., & Gary, T.L. (2007). Have Americans increased their fruit and vegetable intake?: The trends between 1988 and 2002. *American journal of preventive medicine,* vol. 32, pp. 257-263.

Casey, P.H., Szeto, K., Lensing, S., Bogle, M., & Weber, J. (2001). Children in Food-Insufficient, Low-Income Families: Prevalence, Health, and Nutrition Status. *Archives of Pediatrics & Adolescent Medicine,* 155 (4):508-514.

Cole, N., Jacobson, J., Nichols-Barrer, I., Fox, M., & Robare, J.. (2011) WIC Food Packages Policy Options Study. US Department of Agriculture, *Food and Nutrition Service, Office of Research and Analysis. Alexandria,* VA.

Compher, C. (2006). "The Nutrition Transition in American Indians." *Journal of Transcultural Nursing, 17*(3):217-223.

Dietary Guidelines Advisory Ccommittee. (2010). Report of the Dietary Guidelines Advisory Committee on the Dietary Guidelines for Americans, 2010, to the Secretary of Agriculture and the Secretary of Health and Human Services. Agricultural Research Service.

Davis, C &Saltos, E. (1999). Dietary recommendations and how they have changed over time. *America's eating habits: Changes and consequences,* pp. 33-50.

Dong, D., &Lin, B.H.. (2009). *Fruit and vegetable consumption by low-income Americans.* Vol. 70: Citeseer.

Drewnowski, A. &Specter, S. (2004). Poverty and Obesity: The Role of Energy Density and Energy Costs. *The American Journal of Clinical Nutrition,*79(1):6-16.

Dutko, P. (2012). Food Deserts Suffer Persistent Socioeconomic Disadvantage. *Choices,* vol. 27.

Dutko, P., VerPloeg, M., & Farrigan, T.L. (2012). *Characteristics and influential factors of food deserts.* United States Department of Agriculture, Economic Research Service.

Gleason, S. &Pooler, J. (2011). *The Effects of Changes in WIC Food Packages on Redemptions.* United States Department of Agriculture, Food Assistance and Nutrition Research Program.

Ishdorj, A., &Capps, Jr., O. (2013). The Effect of Revised WIC Food Packages on Native American Children. *American Journal of Agricultural Economics,* 95:1266-1272.

Kyureghian, G., Nayga, R.M. & Bhattacharya, S. (2013). The Effect of Food Store Access and Income on Household Purchases of Fruits and Vegetables: A Mixed Effects Analysis. *Applied Economic Perspectives and Policy,* vol. 35, pp. 69-88.

Larson, N.I., Story, M.T., & Nelson, M.C.. (2009). Neighborhood environments: disparities in access to healthy foods in the US. *American journal of preventive medicine,* vol. 36, pp. 74-81. e10.

M. VerPloeg &Breneman, V. (2013). *USDA Economic Research Service -* Food Access Research Atlas..

Norris, T., Vines, P.L., & Hoeffel, E.M. (2012). *The American Indian and Alaska Native Population:* 2010. U.S. Census Bureau. Washington D.C.

O'Connell, M., Buchwald, D.S., & Duncan, G.E.. (2011). Food Access and Cost in American Indian Communities in Washington State. *Journal of the American Dietetic Association,* 111(9):1375-1379.

Rose, D., Bodor, N., Swalm, C., Rice, J., Farley, T., & Hutchinson, P. (2009). *Deserts in New Orleans? Illustrations of urban food access and implications for policy*. Ann Arbor, MI: University of Michigan National Poverty Center/USDA Economic Research Service Research.

Sandefur, G.D., Rindfuss, R.R., & Cohen, B. (1996). *Changing Numbers, Changing Needs: American Indian Demography and Public Health: Introduction,* Washington, D.C.: National Academy Press.

Satia, J. A. (2009). Diet-related disparities: understanding the problem and accelerating solutions," *Journal of the American Dietetic Association,* vol. 109, p. 610.

United States Census Bureau. (2004). Income Stable, Poverty Up, *Numbers of Americans With and Without Health Insurance Rise,* Census Bureau Reports. Washington D.C. U.S. Census Bureau News.

United States Department of Agriculture, Food and Nutrition Service. (2002). *The Characteristics of Native American WIC Participants,* On and Off Reservations. Rep. WIC-02-NAM, Alexandria, VA.

United States Department of Agriculture, Food and Nutrition Service. (2011). *WIC Participant and Program Characteristics* 2010. Rep. WIC-10-PC, Alexandria, VA: December.

United States Department of Health and Human Services, Center for Disease Control. (2005). NHANES 2003–2004. Atlanta, GA: US Department of Health and Human Services.

VerPloeg, M. (2010). Access to affordable, nutritious food is limited in "food deserts." *Amber Waves,* vol. 8, pp. 20-7.

VerPloeg, M. (2010). Food environment, food store access, consumer behavior, and diet. *Choices,* vol. 25.

VerPloeg, M., Breneman, V., Farrigan, T., Hamrick, K., Hopkins, D., Kaufman, P., et al. (2009). Access to affordable and nutritious food: measuring and understanding food deserts and their consequences. Report to Congress. Access to affordable and nutritious food: measuring and understanding food deserts and their consequences. Report to Congress.

Weatherspoon, D. D., VerPloeg, M., & Dutko,P. (2012). An Evaluation of Food Deserts in America. *Choices,* vol. 27.

INDEX

A

B

C

D

E

F

G

H

N

O

P

Q

R

S

T